MILITARY LEGITIMACY

MILITARY LEGITIMACY
Might and Right in the New Millennium

RUDOLPH C. BARNES Jr.

Routledge
Taylor & Francis Group

LONDON AND NEW YORK

First published in 1996 in Great Britain by
Routledge
2 Park Square, Milton Park, Abingdon, Oxfordshire OX14 4RN
711 Third Avenue, New York, NY 10017

First issued in paperback 2016

Routledge is an imprint of the Taylor and Francis Group, an informa business

Excerpts from *A Distant Mirror* by Barbara W. Tuchman
copyright © 1978 by Barbara W. Tuchman.
Reprinted by permission of Alfred A. Knopf Inc.

British Library Cataloguing in Publication data

Barnes, Rudolph C.
    Military Legitimacy: Might and Right in
    the New Millennium
    I. Title
    306.27

Library of Congress Cataloging-in-Publication data

Barnes, Rudolph C., 1942–
    Military legitimacy : might and right in the new millennium /
    Rudolph C. Barnes, Jr.
        p.   cm.
    Includes bibliographical references and index.
    ISBN 0-7146-4624-5 (hardback)
    1. Sociology, Military.   2. Command of troops.   I. Title.
    U21.5.B36   1996
    306.2'7—dc20   95-30804
                                                                CIP

ISBN 13: 978-1-138-98110-2 (pbk)
ISBN 13: 978-0-7146-4624-4 (hbk)

Typeset by Vitaset, Paddock Wood

Publisher's Note
The publisher has gone to great lengths to ensure the quality
of this reprint but points out that some imperfections in
the original may be apparent

To
my family and friends

Blessed be the tie that binds

Contents

Introduction

Military legitimacy is about the balance between might and right. It is a relative concept, differing in periods of war and peace. In wartime, survival takes precedence over the niceties of the law; there can be no substitute for victory. In peacetime the legitimacy of military operations is not measured by overwhelming force but by public support – the vacillating, unwritten product of the public will. The focus of this book is on the legitimacy of peacetime military operations and the public support that is both a requirement and measure of military legitimacy.

Beginning with the United States Constitution, the rule of law and its protection of human rights has been the bedrock of military legitimacy; but there is more to legitimacy than the law. It also includes a shifting penumbra of moral and cultural standards which are interpreted in the context of prevailing values – values that often have different meanings for civilians and military personnel.

National values – the interrelated concepts of democracy, human rights and the rule of law – are ingrained in the legal and moral requirements of military legitimacy. They are the stuff of politics, and politics have traditionally been anathema to the military. Public support is both a requirement and measure of military legitimacy in a democracy.

The subject of military legitimacy is especially relevant to our times. With the end of the Cold War, US national and military strategies, driven by the pervasive threat of the 'evil empire' for 40 years, lost their underpinnings. New strategies and capabilities must be built on the principles of legitimacy, and while the core principles of democracy, human rights and the rule of law have not changed, the new strategic environment is creating new priorities for military legitimacy in the coming millennium.

This book describes the components of military legitimacy and applies them to contemporary military activities. In military doctrine these activities are referred to as operations other than war. Most are civil-military operations which are distinguished from warfighting in

1

that their ultimate objective is not to defeat an enemy with overwhelming force, but to achieve political objectives through public support both at home and in the area of operations.

Civil-military relations are usually an index of success or failure in operations other than war. Achieving mission success in the new millennium will require redefining the role of the soldier and the state to emphasize civil-military relations and strengthening civil-military capabilities through more extensive and effective utilization of civilian soldiers in the total force. This book advocates new paradigms for leadership and civil-military capabilities in operations other than war: they are the diplomat warrior and civil affairs forces.

Conforming military strategies and capabilities to the new priorities of military legitimacy will require change, but change will not come easily to the world's largest bureaucracy. Thomas Jefferson once spoke of the need for change; his words, prominently displayed in the lobby of The Army Judge Advocate General's School, in Charlottesville, Virginia, are especially relevant to military legitimacy:

> Laws and institutions must go hand in hand with the progress of the human mind. As that becomes more developed, more enlightened, as new discoveries are made, new truths disclosed, and manners and opinions change with the change of circumstances, institutions must advance also, and keep pace with the times.

So it is with the US military. As an institution governed by the rule of law it must advance or adjust its course to keep pace with changing times – specifically changing security needs and public perceptions of the role of the military. The focal point for such change will be in military leadership. Military leaders in the new millennium must be more than fighters; they must also be thinkers and diplomats who can serve a broad spectrum of peacetime security needs. More extensive civil-military relations are contemplated by new doctrine on operations other than war. This doctrine brings the military into the mainstream of US domestic and foreign policy, and requires military leaders who can function as an extension of both the military and the diplomatic corps.

Chapter headings suggest a spiritual dimension to traditional values, with religion – specifically the golden rule – providing the moral foundation for the concept of legitimacy. But religious fundamentalism can be a serious threat to democracy, human rights, and the rule of law. Most in the military understand this paradox of religion and democracy, since it is similar to the paradox of the military as an authoritarian organization in a democracy.

Faith is required of those who risk life and limb for God and country; but blind faith and zeal in the military can threaten legitimacy, as it did in Hitler's Germany. Even in the US over-zealous officers, such as Colonel Oliver North, jeopardize military legitimacy whenever they put mission above the law; in a similar manner their kindred religious zealots threaten freedom and democracy whenever they demand that government embrace their religious rules and intolerance for dissent.

History has taught the dangers of misplaced loyalty that comes from military and religious fanaticism. The panoply of faiths that sustain those in the military must have their common denominator in the Constitution, which guarantees the freedoms of religion and expression. Concepts of duty and loyalty must be grounded in preserving and protecting the Constitution and the national values it was created to protect: democracy, human rights and the rule of law.

An understanding of military legitimacy and its relevance to military activities other than warfighting will help future leaders understand the requirements of military professionalism and the dangers of unrestrained zeal in peacetime. It will help future leaders understand the overriding importance of civil-military relations to military legitimacy, and the importance of civilian soldiers to civil-military relations.

But military legitimacy is for civilians as well as those in the military. Civilians are half of the civil-military equation; without broad-based public understanding and support for change, the inertia of the world's largest bureaucracy will carry it into the next century unsuited for the challenges that lie ahead. The changes required are radical, challenging traditional notions of the soldier and the state. This book will have served its purpose if it sparks honest debate on the contentious issues it presents.

Two final notes: because of the controversial nature of the concepts presented, an abundance of authority has been cited to support them; and by way of disclaimer, the author is a civilian soldier who has written this book in his capacity as a civilian, not a soldier. The views and recommendations of the author are his own and not necessarily those of the Department of Defense. In fact, many of the recommendations are likely to be at odds with Pentagon pundits; their implementation will require force-feeding from across the river – from Congress or the President.

Rudolph C. Barnes, Jr.

1
Might and Right, Past and Present

But as regards the towns of those peoples which Yahweh your God gives you as your own inheritance, you must not spare the life of any living thing.

Deuteronomy 20:16

But I tell you, love your enemies ...

Matthew 5:44

Historically the end has justified the means in warfare. Might has made right; victors have set the standards of legitimacy. Even in the Second World War, the Allies were able to set standards at Nuremberg only because they were victorious. And in the light of the wholesale slaughter of civilians at Nagasaki, Hiroshima and Dresden, there is some question whether Allied criteria for crimes against humanity represented a double standard.

Military legitimacy relates to the balance between might and right. Its standards are concerned primarily with protecting civilians and other non-combatants from the ravages of war. Standards of military legitimacy evolved from the Just War tradition and are incorporated in the law of war. But the Judeo-Christian tradition began with holy war, where no distinction is made between combatants and non-combatants, only between good and evil. Its brutal progeny can be seen in contemporary primal violence in the Middle East, Northern Ireland and Bosnia.

The concept of chivalry evolved from the Middle Ages to provide a moral foundation for the protection of civilians in war. A variety of chivalry survived in the antebellum South until it became a cultural casualty of the War Between the States – a war that introduced the theory of collective responsibility to justify total war. In the nineteenth century the US would demonstrate its proficiency in total war, but later learn the painful lessons of limited war.

OLD TESTAMENT HOLY WAR

Old Testament holy war was based on the rule of law – God's law. Then, and now, holy war was a struggle between the forces of good and evil. The ancient Jewish law of war codified in the Book of Deuteronomy made no meaningful distinction between combatants and non-combatants; in holy war ethnic and religious background distinguished friend from foe.

The commitment of the early Jews to the rule of law was more a matter of religious faith than political philosophy. Unlike the founding fathers who wrote the US Constitution and who took special pains to separate the church from the state, ancient Jews saw God as the source of their law and inseparable from military and political events. War was God's way of delivering the Promised Land to His chosen people. For Old Testament Jews, God's wars were Just Wars, and anyone between them and the Promised Land was an enemy who deserved no quarter. The holy end justified any means. With God on their side, might made right.

Chapter 20 of the Book of Deuteronomy sets forth the law regarding treatment of women and children in besieged towns. It makes a distinction between those in far distant towns and those nearby: the former might be taken as slaves as the booty of war,[1] but the latter were to be slaughtered without mercy.[2] Only fruit trees were to be spared the sword, simply because they were not human.[3]

The story of Joshua and the battle of Jericho, reported in Chapter 6 of the Book of Joshua, is an application of the ancient Jewish law of war. After barricading and laying siege to Jericho, 'Yahweh [God] said to Joshua, "Now I am delivering Jericho and its King into your hands"'.[4] After quietly marching around Jericho for six days, on the seventh day a blast of trumpets and a war cry brought down the walls and the victorious Jewish army rushed into the town. 'They enforced a ban on everything in the town: men and women, young and old, even the oxen and sheep and donkeys, massacring them all'.[5]

The mixture of religion and law which characterized Old Testament warfare may seem brutally archaic, but the same Middle East where God first became a warrior over 3,000 years ago remains a hotbed of militant religious fundamentalism. The spread of Islamic fundamentalism and the cry for Islamic holy war (*Jihad*) is based on the same premise as the Jewish holy war that continued from the time of Moses to the creation of a Jewish state in 1948. That premise is that aggressive war is God's way of rewarding His chosen people – a means of divine justice. But that premise is entirely at odds with modern concepts of Just War and human rights.

The merciless dictates of Jewish and Islamic law illustrate the danger of a rule of law without human rights. Christianity provided, in theory at least, a moral alternative. The teachings of Christ, based on *agape* (love), required respect for human dignity and compassion for the suffering, even of one's enemy. For medieval Christian warriors, values such as mercy and *noblesse oblige* were incorporated into the code of conduct known as chivalry. As a legal and moral code, it provided a colorful chapter in the history of military legitimacy. As the essence of honor for the warrior, chivalry remains imbued in the traditional values of the Professional Army Ethic.

CHIVALRY: A DISTANT MIRROR OF MILITARY LEGITIMACY[6]

The noble qualities of knighthood – courage, honor and a readiness to help the weak and protect women – were the substance of chivalry.[7] These qualities were personified in King Arthur and his Knights of the Round Table, whose medieval exploits of good against evil are familiar to every youngster. Unfortunately, history records that the romantic virtues of King Arthur's knights, or at least their real world counterparts, were more myth than reality:

> [They] were supposed, in theory, to serve as defenders of the Faith, upholders of justice, champions of the oppressed. In practise, they were themselves the oppressors, and by the fourteenth century the violence and lawlessness of the men of the sword had become a major agency of disorder.[8]

Chivalry was more than a list of admirable qualities of the medieval warrior. It was a moral system that developed with the great crusades and governed all aspects of life for the Christian nobility of that era. In war, its emphasis was on individual gallantry and honor, making hand-to-hand combat between knights the norm. As a code of conduct, it 'intended to fuse the religious and martial spirits and somehow bring the fighting man into Christian theory'.[9]

The standards of chivalry were virtuous and demanding. Honore Benet, a Benedictine Prior, wrote of them in his *The Tree of Battle*, which he dedicated to Charles VI, in 1387.

> Through every discussion his governing idea was that war should not harm those who do not make war, while every example of his time showed that it did ... With no illusions about chivalry, Benet wrote that some knights were made bold by their desire for glory, others by fear, others by 'greed to gain riches and for no other reason'.[10]

7

Medieval Europe was obsessed with religious war. The Popes, kings and their noble knights saw nothing more virtuous than a crusade to expand their power and influence. Crusades and chivalry dominated fourteenth-century Europe, and coincided with renewed interest in the concept of Just War. For a king to raise and supply an army, the cause had to be just. As with Jewish leaders 1,000 years earlier, Christian kings were convinced that God was on their side when they went to war. But with their rape, pillage and plunder the crusades were no more humane than the holy wars of Old Testament times.

In the fourteenth century kings looked to their knights and vassals to raise the men and money required for war. Unless perceived to be a Just War, the means would not be forthcoming. The elements of Just War have changed little since the fourteenth century: the war had to be declared just by *competent authority*; some injustice on the part of the enemy, like heresy, was needed to ensure a *just cause*; and *right intention* was also required. But none of this moralizing negated the

> ... right of spoil – in practice, pillage – that accompanied a Just War. It rested on the theory that the enemy, being *unjust*, had no right to property, and that booty was the due reward for risk of life in a just cause.[11]

The fourteenth century was a violent time of plagues and persecution. Life was cheap and human rights non-existent. Discrimination against the Jews was especially brutal, their being labelled 'Christ killers' and blamed for the Black Death that ravaged the land. Even Thomas Aquinas opined that 'since Jews are slaves of the Church, she can dispose of their possessions'.[12] With religion the driving force behind medieval politics and warfare, inquisitions and crusades were the norm. They represented a great irony: intolerance, hate and violence in the name of Jesus who taught love, mercy and peace.

The hypocrisy of chivalry was evident in the frequent raids by knights into adjoining lands for the specific purpose of plunder. A classic example was a raid in 1355 by Prince Edward of England across the Channel into Bordeaux:

> Never had the famous, beautiful and rich lands of Armagnac known such destruction as was visited upon it in these two months. The havoc was not purposeless but intended, like military terrorism in any age, to punish or deter people from siding with the enemy. ... Plunder would play its part both as profit and pay. In reporting his raid to the Bishop of Winchester, Prince Edward described it as 'harrying and wasting the country'.[13]

The sanctimonious bubble of chivalry, however, was destined to burst at the battle of Nicopolis in 1396. En route to that fateful battle, French knights reportedly engaged in 'wrongs, robberies, lubricities, and dishonest things', including debaucheries with prostitutes. This indiscipline and misconduct encouraged 'the men in outrages upon the women in countries through which they passed'. Unrestrained pillage and mistreatment of native inhabitants along the way to Nicopolis caused great consternation among clerics who implored the knights to behave, but to no avail. In frustration one monk lamented 'they might as well have talked to a deaf ass'.[14]

When they arrived at Nicopolis on the Danube River in what is now Bulgaria, French knights initiated the battle with a glorious charge against the Turkish infidels holding the city. Because of poor discipline, non-existent strategy and worse tactics, the crusaders, while initially successful, were routed by the better led and disciplined Turkish forces of Sultan Bajazet and his Bulgarian allies. The bloody defeat ended the Crusades and the era of the gallant knight.[15] The battle of Agincourt in 1415 confirmed the end of chivalry as operational art. At Agincourt, as at Nicopolis, 'the battle was lost by the incompetence of French chivalry, and won more by the action of the English common soldiers than of the mounted knights'.[16]

While chivalry became obsolete as an operational concept, it has survived in military concepts of honor and gallantry which incorporate the moral imperative to protect civilians from the ravages of war. During the War Between the States chivalry flourished among officers of the Confederate army. It was personified in the dignity of General Robert E. Lee and the more colorful exploits of General J. E. B. Stuart. And while chivalry has produced a sometimes distorted sense of honor, its ideals continue to imbue the traditional military values of duty, loyalty, integrity and selfless service.

According to Lewis Lapham, chivalry is alive and well; President Bush relied on it to provide legitimacy to *Restore Hope*, the 1992 humanitarian intervention in Somalia:

> Because the American public likes to believe that its cause is either noble or just, the argument must be phrased in the language of chivalry or Holy Crusade, and when President Bush sent the army to Somalia last December he borrowed the persona of a medieval pope.[17]

TOTAL WAR AT HOME: THE BURNING OF COLUMBIA

In 1860 it was not chivalry but a commitment to preserve the Union that motivated President Lincoln to go to war to prevent the secession of the Confederate states of America. At the height of that war in 1863 the US adopted the Lieber Code as General Order No. 100. It was a landmark statement of military legitimacy and civil-military relations that confirmed a principle at the heart of the law of war: those who do not make war should be protected from its harm.[18]

The Code was written by Francis Lieber, who emigrated to the US from Germany in 1827 after being imprisoned by Prussian police on suspicion of being a revolutionary. He settled in the deep South, assuming a professorship at South Carolina College (now the University of South Carolina) in Columbia, South Carolina. Lieber left the city of Columbia for Columbia University in the 1850s, during a time of political intolerance when southern 'fire-eaters' effectively purged many intellectuals who did not embrace their views, including the need to maintain the 'peculiar institution' of slavery.

Professor Lieber could not have known that his adopted city, Columbia, would be destroyed by the Union army in 1865 in violation of his Code. The provisions of the Lieber Code then governed military operations as US law, and would become international law when incorporated in the Hague Conventions of 1899 and 1907, and the Geneva Conventions of 1949. During the War Between the States there were violations of the Lieber Code on both sides; but with the exception of partisans who were beyond the control of regular commanders, violations were rarely egregious and were usually denounced by senior commanders.

On the Confederate side, General Robert E. Lee exemplified the ideals of chivalry when he moved his army into Maryland and Pennsylvania. The citizens of these states remarked at the perfect discipline of Lee's rag-tag rebels as they marched past their homes. This was the result of instructions given by Lee to his men that reflected his moral conviction that civilians should be spared the ravages of war:

> I cannot hope that heaven will prosper our cause when we are violating its laws. I shall therefore carry on the war in Pennsylvania without offending the sanctions of a high civilization and Christianity.[19]

Lee treated civilian property with respect. Rather than have his men live off the land, he instructed his commissary officers to make formal requisitions when supplies were needed. Lee made a distinction between combatants and non-combatants when he exhorted his troops

> ... to abstain with most scrupulous care from unnecessary or wanton injury to private property. It must be remembered that we make war only upon armed men.[20]

The respect accorded enemy civilians by Lee was in stark contrast to the scorched earth strategy of Union General William Tecumseh Sherman, who had been given the mission of destroying Confederate forces in the deep South while Grant hammered Lee in northern Virginia. General Sherman did not share the philosophy of Lee, nor did his tactics reflect even a hint of chivalry. In fact, while Sherman gave lip service to the Lieber Code, his troops consistently violated its provisions.

Sherman was an advocate of total war, having declared his philosophy as early as October 1862. Total war was based on collective responsibility, which allowed for little real distinction between combatants and non-combatants. Sherman believed the Union must 'make the war so terrible' for all rebellious Southerners that they would never again revolt. To accomplish this, the Southerners must 'be made to fear us, and dread the passage of our troops through their country'.[21]

General Sherman's views may have been influenced by Prince Edward's raid into France in 1355; harrying and wasting the South was Sherman's stated objective. After burning Atlanta to the ground, it did not take long for Sherman's men to get the hang of plunder and pillage. By the time they reached Savannah they had destroyed vast areas of the Georgia heartland. But it was just a preview of what awaited the Carolinas.

In January 1865, Sherman's 60,000 veteran troops moved out of Savannah, made a feint toward Charleston and Augusta, and then moved toward Orangeburg and Columbia. Sherman left no doubt that he intended to punish the South Carolinians, as theirs was the first state to secede, and make a special example of Columbia, since the act of secession had taken place there.[22]

In Savannah, Sherman had promised vengeance for the Union in South Carolina: 'I look upon Columbia as quite as bad as Charleston, and I doubt we shall spare the public buildings there as we did in Milledgeville'. He also acknowledged the hatred among his men for the Palmetto state: 'The truth is the whole army is burning with an insatiable desire to wreak vengeance upon South Carolina. I almost tremble for her fate, but feel that she deserves all that seems in store for her.'[23]

Along the way to Columbia, Sherman's men demonstrated their talent for plunder and pillage, not to mention wanton destruction. In Hardeeville, a church was destroyed piece by piece, with soldiers heckling local residents as the church collapsed.[24] 'Bummers' were the primary vandals: they were soldiers who did their own thing, but were seldom disciplined for their indiscretions. They were especially fond of destroying pianos with their hatchets, competing to see who could

11

make the most noise, breaking dishes, and dressing up in the finest women's clothes.[25]

Sherman's army arrived on the banks of the Congaree River opposite Columbia on 16 February 1865, and the next day the Mayor of Columbia, T. J. Goodwyn, surrendered the defenseless city to Sherman. The general and his staff spent the afternoon with notables, but the troops had their own priorities. They arrived singing 'Hail Columbia, happy land; if I don't burn you, I'll be damned'.[26]

The city was awash with liquor, and friendly house slaves were passing it out to the feisty troops as they began their looting sprees. Soon things were out of control, whether by design or accident, and by evening drunken soldiers were torching everything that would burn. There was no doubt that the fires were intentionally started by Union troops. Some hurried from block to block carrying wads of turpentine-soaked cotton, while others interfered with firefighting efforts. The only issue was whether Sherman authorized the destruction or not; he vehemently denied having done so. He initially blamed the mayor for the free-flowing liquor, citing the impossibility of controlling his drunken soldiers, but he later blamed General Wade Hampton, a popular native son whose cavalry had been the last Confederate troops to leave Columbia.[27]

By morning, 84 of the 124 blocks of Columbia had been burned. Included in the destruction were churches, an Ursuline convent, all public buildings except the unfinished statehouse, as well as most of the city's private residences, of rich and poor alike. Sherman's reaction, other than disclaiming responsibility, was the rationale of collective responsibility:

> Though I never ordered it, and never wished it, I have never shed any tears over it, because I believe it hastened what we all fought for – the end of the war.[28]

The Union troops did more than burn the city. There were many reported violations of human dignity, if not assaults, upon the women of Columbia.

> An extreme practise followed by a few soldiers in looking for valuables hidden on a woman's person was to catch her by the throat and feel in her bosom for a watch or pull up her dress in search of a purse hidden in her girdle or petticoat. Those not so brazen did not hesitate to point a pistol at a woman's head to learn the location of the family heirlooms.[29]

While there were few reported cases of rape against white women, the same was not true for black women. On the morning of 18 February 'their unclothed bodies, bearing the marks of detestable sex crimes, were found about the city'.[30]

After 17 February, pillage and plunder became more restrained. But the soldiers never showed any repentance for their acts, and

> made no pretense of hiding their loot. Stolen jewelry and coin were very much in evidence on their persons as they strolled the streets boasting of having burned Columbia.[31]

When Sherman's men finally left Columbia on 20 February, they had earned the lasting enmity of the people of Columbia, the South, and even some Yankees:

> Whitlaw Reid, the Ohio politician, called the burning of Columbia 'the most monstrous barbarity of the barbarous march'. The people of Columbia, in full agreement with Reid, were also positive that one day the Devil 'with wild sardonic grin, will point exultant to a crime which won the prize from SIN'.[32]

The war ended at Appamattox later that year. Sherman held to his belief that his punishment of southern civilians contributed to Lee's surrender, although there is little evidence to that effect. To the contrary, Sherman's total war tactics created a hatred for him and the Union that made relations between the North and South difficult for many years. It was a legacy of hate that would take more than a century to heal, and which would never be forgotten.

TOTAL WAR IN THE TWENTIETH CENTURY

While the illegality and impracticality of Sherman's total war strategy may seem obvious, 80 years after he burned Columbia the US resorted to it once again. Even with the outcome of the Second World War a foregone conclusion, the US fire-bombed Tokyo and Dresden and then used the atomic bomb on Hiroshima and Nagasaki. The objective was the same as in the War Between the States: to hasten the end of the war and thereby minimize US casualties.

In the black humor of his classic novel, *Slaughterhouse Five*, Kurt Vonnegut speaks through his discombobulated protagonist, Billy Pilgrim, as the conscience of a country that has buried its sins. Vonnegut, a fourth-generation German-American, was an American Infantry Scout who witnessed the fire-bombing of Dresden as a prisoner-of-war. He applied his considerable literary talent to tell the tale through the wacky Billy Pilgrim.

Ironically, Billy (alias Vonnegut) and other US prisoners were kept in an underground slaughterhouse during the night bombing raid on Dresden, 13 February 1945, so that they survived the holocaust while over 130,000 Germans were incinerated above. When they emerged from their sanctuary the next day, Billy noted the complete devastation, with 'little logs lying around. They were the people who had been caught in the fire storm'.[33]

A similar horror took place at Hiroshima later that year; but in this case there was but one explosive device, and it was nuclear. The devastation came out of an otherwise quiet sky on the morning of 6 August 1945. John Hersey has captured the horror and heroics of those on the scene that fateful day in his classic *Hiroshima*.[34] Through the eyes of six survivors, Hersey describes how over 100,000 Japanese civilians were killed by the blast in Hiroshima alone, while countless others were destined to suffer and later die from the lingering, mysterious effects of radiation poisoning. Then came Nagasaki, with similar results.

Like Sherman's devastation, US policy-makers justified the mass slaughter at Dresden, Hiroshima and Nagasaki on the theory of collective responsibility: that civilians share the blame for war and should not be spared its suffering. The concept was and remains in violation of the law of war as first codified in the Lieber Code and later affirmed in the Geneva Conventions: civilians are non-combatants and cannot be targets in war.

The 1945 devastation was different from that of 80 years earlier in an important way: it involved the impersonal high-tech weaponry of mass destruction directed against civilian targets. Coming from a silent sky, the nuclear explosion over Nagasaki seemed unrelated to the conventional violence of warfare; for many of its victims it had the character of divine retribution. But the carnage caused by airborne weapons of mass destruction against non-military targets made total war strategy in the twentieth century even more immoral than Sherman's hands-on scorched earth strategy.

THE AMBIGUITY OF LIMITED WAR

Twenty years after the bombing of Nagasaki, US Marines were dispatched to Vietnam – the first contingent of US combat forces to what was to become America's worst military and political disaster. Vietnam was a limited, not a total war; but saturation bombing strategies coupled with endemic ambiguity in distinguishing combatants from non-combatants provided a double standard of military legitimacy for US soldiers.

At the time, the provisions of the Geneva Conventions of 1949 provided the core of the law of war. The Geneva Conventions expanded upon the principles of the Lieber Code: only combatants were lawful targets; civilians were entitled to protection from violence. More restrictive rules of engagement (ROE) embellished the law of war with changing and often irrational limitations on the use of force. The result was confusion and frustration, sometimes accompanied by tragic results.

One of the Marines who arrived in Vietnam in 1965 was Second Lieutenant Philip Caputo. His book, *A Rumor of War*, recounts the ambiguities that undermined military legitimacy in that violent and unforgiving environment. Caputo arrived in Vietnam a highly motivated Marine officer. His collective values were typical of other young officers: he was serving his country, and wanted more than anything else to please the authorities who judged him. Caputo cited the story of Jesus and the centurion in *Matthew 8:9* to illustrate the timeless nature of military authority. He recalled an incident when he was chewed out for smoking during a tactical operation. After that traumatic experience, he 'turned into a regular little martinet'. Looking back, Caputo felt that experience shaped his attitude in Vietnam:

> Napoleon once said that he could make men die for little pieces of ribbon. By the time the battalion left for Vietnam, I was ready to die for considerably less, for a few favorable remarks in a fitness report.[35]

As motivated as he was, it did not take long for the ambiguities of Vietnam to cause Caputo to question the infallibility of military authority. ROE which incorporated the legal requirements of the Geneva Conventions and shifting US policy did not clarify murky issues of military legitimacy:

> The day before a rifleman in B company had shot a farmer, apparently mistaking him for a VC. To avoid similar incidents in the future, brigade again ordered that chambers be kept clear except when contact was imminent, and in guerrilla-controlled areas, no fire be directed at unarmed Vietnamese *unless they were running*. A running Vietnamese was a fair target. This left us bewildered and uneasy. No one was eager to shoot civilians.

After many questions from his men, 'the skipper finally said, "Look, I don't know what this is supposed to mean, but I talked to battalion and they said as far as they're concerned, if he's dead and Vietnamese, he's VC." And on that note we left to brief the squad leaders.'[36]

The convoluted logic of ROE created moral and ethical dilemmas for Caputo and his fellow Marines:

It was morally right to shoot an unarmed Vietnamese who was running, but wrong to shoot one who was standing or walking; it was wrong to shoot an enemy prisoner at close range, but right for a sniper at long range to kill an enemy soldier who was no more able than a prisoner to defend himself; it was wrong for an infantryman to destroy a village with white-phosphorus grenades, but right for a fighter pilot to drop napalm on it.[37]

Caputo recalled the ethical issues of bombing civilian targets in the Second World War when he talked of the hypocrisy of condoning high-tech massacres in Vietnam. To him and his cohorts it appeared distance from the target reduced the culpability of those who delivered lethal force:

Ethics seemed to be a matter of distance and technology. You could never go wrong if you killed people at long range with sophisticated weapons.[38]

But even with his doubts Caputo remained the quintessential Marine who asked few questions and accepted whatever answers authority gave him. He sounded eerily like a young Oliver North. He recalls:

In the patriotic fervor of the Kennedy years, we had asked, 'What can we do for our country?' and our country had answered, 'Kill VC'. That was the strategy, the best our military minds could come up with: organized butchery. But organized or not, butchery was butchery, so who was to speak of rules and ethics in a war that had none?[39]

The ambiguity and illegitimacy of Vietnam got personal for Caputo when he had a month still to go on his tour. Stung by criticism that his men had mistakenly released two VC during a raid, Caputo directed his men on a manhunt for them. Two Vietnamese were captured, but both were killed en route to headquarters. When the bodies were examined it was determined that they were not VC but innocent civilians. For this mistaken identity Caputo and his men were to be tried by general court-martial.[40]

Thanks to an aggressive Marine Corps defense counsel, the first defendant to be tried was acquitted and charges dropped against the others. Caputo had his own explanation for the result, which illustrated the injustice of an ambiguous war:

The killings had occurred in war. They had occurred, moreover, in a war whose sole aim was to kill Viet Cong, a war in which those ordered to do the killing often could not distinguish the Viet Cong from civilians, a war in which civilians in 'free-fire zones' were killed every day by weapons far more horrible than pistols or shotguns. The deaths ... could not be divorced from the nature and conduct of the war. They were an inevitable product of the war. As I had come to see it, America could not intervene in a people's war without killing some of the people.[41]

The ambiguity of the Vietnam conflict made it difficult, if not impossible, to make a distinction between combatant and non-combatant – the distinction upon which the legitimate use of lethal force depends. And its ROE illustrated a double standard for killing civilians that was left over from the Second World War: it was right to kill large concentrations of civilians with strategic bombing, but wrong to kill them individually with tactical weapons. Ultimately these and other moral and political issues eroded the public support required to sustain US involvement in Vietnam, resulting in an ignoble withdrawal in 1975.

THE 100-HOUR WAR

In *Desert Storm* there was little ambiguity and even less resistance, thanks to Saddam Hussein, who proved to be a classic and inept villain. In support of UN Resolutions, Congress gave the President *carte blanche* to push Hussein out of Kuwait, and US General Norman Schwarzkopf was the modern-day John Wayne who put the villain in his place.

In the 100-hour blitzkrieg and the aerial bombardment that preceded it, the law of war and ROE provided standards of legitimacy for US forces. And while most commentators considered the bombing to be in accordance with the standards of Just War, for some the targeting of infrastructure (for example, the destruction of utilities serving civilian as well as military needs) raised the issue of excessive force.[42]

Desert Storm was a limited war fought during peacetime. It illustrated the mix of legal and political issues which require moral choice. One such issue involved targeting Scud missiles:

Although going after them at night is justifiable under the doctrine of military necessity ... the question did not stop there. 'Once a commander realized civilians were being injured when we went after the launchers, he had to weigh the issues', says Hays Parks, special assistant for law of war matters in The Office of the Judge Advocate General of the Army. 'He's going to ask the JAG [military lawyer], can I do it? And the JAG will say, yes, you can. But any good commander, and especially one as smart as Schwarzkopf, will ask, is this any way to fight a war?' Parks says. 'It underlines the fact that few legal decisions are made in a vacuum. They are often fraught with political and moral considerations.'[43]

Even with its success, *Desert Storm* left questions relating to military legitimacy unanswered:

> For example, now that we have demonstrated our ability to hit targets with pinpoint accuracy, isn't it possible that any failure to do so (and so minimize collateral damage to civilians and surrounding buildings) will invite charges of indiscriminate bombing and be deemed an infliction of unnecessary suffering and a violation of the law of war?[44]

The answer is yes, the destructive power and surgical capability of modern weaponry place a premium on the competence of operators using them. General Schwarzkopf spoke often about the need to avoid collateral damage, but then cited General Sherman as one of his heroes. While Schwarzkopf shared Sherman's well-known hatred for war and feisty determination to end it quickly, he did not appear to embrace Sherman's total war strategy. It was an anomaly noted by at least one military writer:

> Whether judged in the light of tactics or moral conduct, the actions of the American military in the Gulf War reflected the impact of [General Robert E.] Lee, not Sherman.[45]

But modern strategists have noted that General Sherman's interpretation of collective responsibility is alive and well. Quincy Wright has theorized that total war 'caused a breakdown of the distinction between the armed forces and the civilians in military operations', so that civilians 'can no longer expect to be exempt from attack.'[46] Similarly Elliot Cohen has supported Sherman's rationale for punishing civilians in war and applied it to strategic bombing:

> In many cases today, war means bringing power, particularly air power, to bear against civilian society.
>
> The electric generators that keep a defense ministry's computers running and its radars sweeping the skies also provide the energy for hospitals and water purification plants. The bridges indispensable to the movement of military forces support the traffic in food, medicine and all other elements of modern life for large civilian populations. Sherman, reflecting the character of armed struggle in his century as well as ours, believed that in modern conditions civil society must inevitably become a target.[47]

Justifying civilians as strategic targets while prohibiting them as tactical targets is an untenable double standard, as illustrated by the frustration of Philip Caputo in Vietnam. Joseph Sobran has noted the moral, if not legal, hypocrisy of targeting civilians. Southerners once understood the moral implications based on personal experience with Sherman, and Germans and Japanese still remember that two-thirds of those killed in the Second World War were civilians; but the issue is seldom raised. Sobran suggests that ugly wartime memories are reprogrammed to conform to more acceptable humanitarian images.

> Every war becomes humane in retrospect. The victors in every war feel compelled to portray themselves not only as powerful but as humanitarian. The Civil War is now primarily remembered as a war to end slavery rather than a war to prevent, and punish, secession – though freeing the Confederacy's slaves was a punitive expropriation. Despite the US government's campaign of hatred against the Japanese, World War II is similarly remembered as a war against racism. Even the nuking of Hiroshima is defended as humanitarian: It 'shortened the war' and saved countless lives.
> See how the modern state has warped our sense of right and wrong? We have become such people as would have appalled our ancestors.[48]

Martin van Creveld has also noted the hypocrisy of the Allies targeting civilians in the Second World War and then punishing Axis leaders for violating the laws of war. But looking ahead to more ambiguous low-intensity conflict, Creveld has predicted that their strategy '... will focus on obliterating the existing line between those who fight and those who watch, pay, and suffer.'[49]

Wright, Cohen, Sobran and Creveld confirm the continuing relevance of the legal and moral dilemma central to military legitimacy and civil-military relations in increasingly ambiguous environments: how to balance the need to achieve military objectives with political objectives that require public support. Collateral damage was not an issue in *Desert Storm*, since public support in the area of operations was not an objective. But *Desert Storm* is the exception that will likely prove the rule: in most contemporary strategic environments public support will be essential to US political objectives.

MILITARY LEGITIMACY AND ANARCHY

If *Desert Storm* demonstrated the effective use of combat power, *Restore Hope* in Somalia reminded US strategists of Vietnam and the limitations of combat power. Operation *Restore Hope* was initiated in

19

December 1991 to provide security for humanitarian assistance in Somalia. It was unique in that there was no functioning government to raise issues of sovereignty; US forces moved into an environment of famine aggravated by anarchy.

Restore Hope was only the first phase of humanitarian assistance to Somalia; largely because political and military objectives were reasonably clear and understood by all concerned the first phase ended on a successful note in May 1993 when the US relinquished operational control of all remaining forces to the UN. In the second phase, *UNISOM II*, it was no longer clear whether the mission was humanitarian assistance or political warfare with uncooperative clans; there was no end state, no government to support, and no peace to keep. In the midst of uncertainty and ambiguity US forces remained until mission creep eroded restraint.

In the primal violence that characterized Somalia, UN forces were constantly under siege from competing warlords. General Aideed, the strongest of the warlords, was thought to be responsible for an ambush that killed a number of UN forces in the summer of 1993. In an effort to capture General Aideed, in October 1993 the US mounted a search and destroy mission against Aideed in Mogadishu. The raid met unexpected resistance and reinforcements were too late to prevent disaster.

The legitimacy of the US presence in Somalia was lost when the American public witnessed televised images of an angry mob of Somalis dragging a dead US soldier through the streets following the abortive Mogadishu raid. The resulting public outcry caused President Clinton to pledge the early withdrawal of US troops and gave General Aideed undeserved legitimacy among Somalis at the expense of UN nation-building efforts. The strategic error of the US was to allow mission creep to convert a defensive humanitarian assistance mission into an offensive peace enforcement mission without fully understanding the implications.

Somalia also provided examples of how too little force can be as fatal to legitimacy as too much force. Feuding warlords forced US and UN forces to remain barricaded and of little use in disarming the roving bandits that terrorized the country. Law and order remained elusive in Somalia in spite of the US and UN military presence. Any such environment requires the use of force to establish law and order, but not offensive combat operations such as the abortive raid in Mogadishu.

RELIGION: THE MORAL FOUNDATION OF LEGITIMACY

The intractable anarchy in Somalia is representative of the primal conflict permeating much of the Third World. One strategist has suggested that religion may be the only antidote for anarchy:

> The key to a positive future in the Third World would be the rise and rapid spread of an alternative value framework and politico-economic system stressing population control, ecological sanity, intergroup cooperation, and deference to authority. Because of the extent of change needed and the speed with which it must take place, such a new system would probably have to take the form of a unifying religion, either a totally new one or a mutation from an existing one. Only a religion can generate the transformative power needed to change the course of the Third World future.[50]

If religion is a cure for anarchy, it is a cure that can be worse than the disease. Fundamentalist religion has historically been a threat to democracy, human rights, and the rule of law. This chapter began with the ancient role of religion and war and has concluded on the same subject, but with a different twist: the idea that religion provides not only the motivation for violence (holy war), but, as the moral foundation of legitimacy, religion also provides the best hope for a lasting peace.

Vaclav Haval, who as president of the Czech Republic has personally experienced the violence of runaway self-determinism, was in Philadelphia to celebrate democracy and receive the Liberty Medal on the Fourth of July, 1994. He spoke of the need for religion in a 'disconnected, chaotic and confusing world', describing it as a 'forgotten awareness' that must be related to human rights:

> This forgotten awareness is encoded in all religions. All cultures anticipate it in various forms. It is one of the things that form the basis of man's understanding of himself, of his place in the world, and ultimately of the world as such ... Politicians at international forums may reiterate a thousand times that the basis of the new world order must be on universal human rights ... [but] only someone who submits to the authority of the universal order ... can genuinely value himself and his neighbors, and thus honor their rights as well.[51]

No American leader had a better understanding of the moral dimension of legitimacy than Thomas Jefferson. As the author of the Declaration of Independence and the First Amendment to the Constitution, Jefferson exemplified the importance of religious faith to the integrity and selfless service needed by all public leaders; but he also understood that the volatile mix of religion and politics required a separation between church and state to preserve religious and political freedom.

Jefferson's understanding of the moral dimension of legitimacy is preserved in a book he prepared for his personal use entitled *The Morals of Jesus*. It was distilled from the teachings of Jesus which Jefferson considered '... the sublimest morality that has ever been taught'. But while Jefferson embraced the teachings of Jesus as a moral code, he detested the way the church had converted these teachings into religious doctrine as '... corruptions of it which have been invested by priestcraft and kingcraft'.[52]

Jesus taught the moral imperatives of *agape* (love) which are at the core of the great religions of the world: Judaism and Islam as well as Christianity. These moral teachings can be summarized in the golden rule: *do unto others as you would have them do unto you*. Biblical scholars disagree over whether this saying originated with Jesus or his contemporary, Rabbi Hillel. All would agree that it is more than a Judeo-Christian concept: it is a universal moral principle upon which altruistic adherents of all the world's religions can agree. And finding common ground among these often contentious religions is imperative to living together in peace.

The golden rule is at the foundation of the Just War Tradition and the Constitution; and as such is part of the spirit of our laws. American presidents have cited the golden rule to be at the core of US national values and foreign policy. It is also at the heart of the Professional Army Ethic in the personal values of integrity and selfless service. These values have real relevance to military and political legitimacy in a world beset by primal violence. The religious and ethnic conflict at the heart of contemporary violence is based on intolerance, fear, and hate, for which the golden rule is the best antidote:

> Fear generates hate, hate provokes counterhate, which in turn creates more fear – until fear and hate both explode into war, in our time truly the war to end all war, for there will be no one left capable of fighting World War IV.[53]

Thomas Jefferson's legacy for legitimacy was a universal moral code based on the teachings of Jesus and a rule of law that would protect human rights from religious intolerance. At the same time he understood that intolerant and coercive religions were the enemy of democracy, human rights, and the rule of law.

Karl Marx once referred to religion as the opiate of the masses. That it is, but it is the kind of sedative required if order is to be restored from the chaos of primal conflict and anarchy. Military forces are no exception; to accomplish their missions in such unforgiving environments they must be sustained by faith[54] as well as professional competence and diplomacy. To promote effectively democracy, human rights, and the rule of law in such environments, US personnel must treat others as they would expect to be treated – such is the spirit of the law and of military legitimacy.[55]

SUMMARY

History has provided examples of holy war, total war, limited war, and a variety of military operations other than war to illustrate the relative importance of civil-military relations to military legitimacy. In unlimited wars, might has most often made right, with civilians little more than obstacles to combat operations. Conversely, in limited wars and operations other than war, might has most often had to be right to be successful, and civilian support has often been the difference between success and failure. The *right* of the equation is what military legitimacy is all about; and concepts of what is right, or legitimate, for the military is the evolutionary product of religious, moral, and legal principles and values that have been inextricably woven into the US national fabric.

Religion – specifically the Judeo-Christian tradition – has had a major influence on US perceptions of military legitimacy, including concepts evolved from holy war and chivalry. The most important principle derived from the code of chivalry was that civilians who do not make war should not suffer from it. A derivative of the golden rule, it provides the moral foundation for humanitarian standards governing military operations. For obvious reasons it cannot be applied to warring combatants whose purpose is to destroy one another; but it can and should be applied to non-combatants. This principle was an integral part of the 1863 Lieber Code and the 1949 Geneva Conventions which are at the core of the law of war; but even the US has failed to apply the principle uniformly.

The following chapters build on experience from the past to develop the strategies and capabilities needed for military legitimacy and leadership in the new millennium. Looking ahead, the focus is not on war but on operations other than war – not on combatants but on non-combatants. Military legitimacy is a concept that is dramatically different in war and peace; and in the ambiguous and unforgiving environments of the new strategic environment it can be the difference between military victory and political defeat. This must be understood by all who wear the uniform and those who would put them in harm's way.

NOTES

1. *Book of Deuteronomy*, Chapter 20, verses 12–14. Martin van Creveld has described Old Testament Holy War (*milchemet mitzvah*) as 'a war of extermination in the fullest sense of that term'. Creveld, *The Transformation of War* (New York: The Free Press, 1991), p. 135.
2. Ibid., verses 15, 16.
3. Ibid., verses 10, 19.
4. *Book of Joshua*, Chapter 6, verse 2.
5. Ibid., verse 21.
6. The reference is to Barbara W. Tuchman's, *A Distant Mirror* (New York: Ballantine Books, 1978).
7. *Webster's New World Dictionary, Third College Edition* (New York: Webster's New World, 1988).
8. Tuchman, *A Distant Mirror*, supra n. 6, Foreword, p. xix.
9. 'Since a knight's usual activities were as much at odds with Christian theory as a merchant's, a moral gloss was needed that would allow the church to tolerate the warriors in good conscience and the warriors to pursue their own values in spiritual comfort. With the help of Benedictine thinkers, a code evolved that put the knight's sword arm in the service, theoretically, of justice, right, piety, the Church, the widow, the orphan and the oppressed.' Ibid, p. 62.
10. Benet was also 'heart-stricken to see and hear of the misery inflicted upon poor laborers ... through whom, under God, the Pope and all the kings and lords in the world would have their meat and all their drink and clothing.' He also stated that it was not permissible to take prisoner the 'merchants, tillers of the soil, and shepherds of the enemy.' Ibid, pp. 414, 415.
11. Ibid., p. 73.
12. Ibid., p. 110.
13. Ibid., pp. 137–38.
14. Ibid., p. 553.
15. Ibid., p. 563.
16. Ibid., pp. 583–84.
17. Lewis H. Lapham, 'Notebook: God's Gunboats', *Harper's Magazine*, February 1993, p. 10.
18. The Lieber Code defines the limits of military necessity by requiring a distinction be made between combatants as lawful targets and non-combatants, who should be protected from unnecessary suffering. Martin van Creveld has referred to The Lieber Code as 'the Union text on international law' and decreed that the rebels would be treated as if engaged in an international conflict. See Creveld, *The Transformation of War*, supra n. 1, p. 41. The following provisions of The Lieber Code were applicable during the War Between the States and remain standards of military legitimacy today:

    Article 15 describes military necessity as allowing the destruction of '*armed* enemies, and of other persons whose destruction is incidentally *unavoidable* in the armed contests of the war' (emphasis in the original). *Article 16* goes on to provide that 'military necessity does not admit of cruelty'.

    Article 22 makes the critical distinction between soldier and civilian, 'between the private individual belonging to a hostile country and the hostile country itself, with its men in arms. The principle has been more and more acknowledged that the unarmed citizen is to be spared in person, property, and honor as much as the exigencies of war will admit.' *Article 23* expands on this theme by ensuring that 'the inoffensive individual is as little disturbed in his private relations as the commander of the hostile troops can afford to grant in the overruling demands of a vigorous war'.

Articles 24 and *25* compare the practise of barbarous armies with that of Europeans and their descendants. 'The almost universal rule in remote times was, and continues to be with barbarous armies, that the private individual of the hostile country is destined to suffer every privation of liberty . . .' In contrast, 'In modern regular wars of the Europeans, and their descendants in other portions of the globe, protection of the inoffensive citizen is the rule.'

Article 38 prohibits the seizure of private property except for military necessity. *Article 42* declares slavery to be against the law of nature, citing Roman law to the effect that 'so far as the law of nature is concerned, all men are equal.' *Article 43* requires that any slave that comes into the hands of US forces be treated as a free person under the shield of the law of nations.

Under *Article 44*: 'All wanton violence against persons in the invaded country, destruction of property not commanded by the authorized officer, all robbery, all pillage or sacking, even after taking a place by main force, all rape, wounding, maiming, or killing of such inhabitants, are prohibited under the penalty of death.'

Article 46 goes further, making it unlawful for US officers or soldiers in a hostile country to make use of their military position for private gain, and *Article 47* makes all common crimes punishable against US forces in a hostile country.

Article 155 requires that the distinction between combatants and non-combatants in regular war be applied to a government in rebellion, and that military commanders protect loyal citizens. Disloyal citizens can be made to bear the burden of war, but this does not authorize violations of their rights under previous provisions.

The above provisions are part of the 157 Articles of The Lieber Code, General Order No. 100, 24 April 1863, published in *The Military Laws of the United States*, War Department Document No. 64 (Washington: Government Printing Office, 1897), pp. 779–799. They have been incorporated in the *1949 Geneva Convention Relative to the Protection of Civilians in Time of War*, which is set forth in FM 27–10, *The Law of Land Warfare* (July 1956). But there is ambiguity over the treatment of enemy civilians in this Army FM; see discussion in note 20 to Chapter 4, infra.

19. Shelby Foote, *The Civil War, Volume 2: Fredericksburg to Meridian* (New York: Random House, 1986), p. 444 (hereinafter *The Civil War*).

20. Lee also told his men 'we cannot take vengeance for the wrongs our people have suffered without lowering ourselves in the eyes of all whose abhorrence has been excited by the atrocities of our enemies . . .' Idem. The *Operational Law Handbook* (JA 422), Center for Law and Military Operations and International Law Division, The Judge Advocate General's School, Charlottesville, VA (1993), cites the following quotes from General Lee and General Sherman as a contrast in command:

> No greater disgrace can befall the army and through it our whole people, than the perpetration of barbarous outrages upon the innocent and defenseless. Such proceedings not only disgrace the perpetrators and all connected to them, but are subversive of the discipline and efficiency of the army, and destructive of the ends of our

movement . . . [T]he duties exacted of us by civilization and Christianity are not less obligatory in the country of the enemy than in our own.
(General Lee on marching into Pennsylvania.)

I sincerely believe that the whole United States, North and South, would rejoice to have this army turned loose on South Carolina, to devastate that state in the manner we have done in Georgia.
(General Sherman on marching into South Carolina.)

21. John G. Barrett, *Sherman's March Through the Carolinas* (Chapel Hill: University of North Carolina Press, 1956), pp. 15–16 (hereinafter *Sherman's March*).
22. Ibid., p. 38.
23. *The Civil War*, pp. 753–54.
24. *Sherman's March*, p. 53.
25. Ibid., p. 55.
26. Ibid., p. 75.
27. *Sherman's March*, pp. 81–90; *The Civil War*, pp. 793–96.
28. *Sherman's March*, p. 91; *The Civil War*, p. 795.
29. *Sherman's March*, p. 85.
30. Idem.
31. Ibid., p. 92.
32. Ibid., p. 89.
33. Kurt Vonnegut, *Slaughterhouse Five* (New York: Dell Publishing, 1968), p. 179.
34. John Hersey, *Hiroshima* (New York: Bantam Books, 1986). The controversy over the *Enola Gay* exhibit at the Smithsonian on the fiftieth anniversary of the bombing of Hiroshima indicated that there is not yet a national consensus on its legality or morality. But John Hersey and The Lieber Code (see n. 18, supra) were cited to support the conclusion that the bombing of Hiroshima and Nagasaki could not be justified under the principles of military necessity, unnecessary suffering, and proportionality. See Maritz Ryan, *The Atomic Bombing of Japan: Military Necessity, Unnecessary Suffering, and Proportionality*, unpublished paper #43015, Center for Law and Military Operations, The Judge Advocate General's School, Charlottesville, VA, 1995. See also n. 4 to Chapter 3, infra.
35. Philip Caputo, *A Rumor of War* (New York: Ballantine Books, 1986), p. 33.
36. Ibid., p. 69.
37. Ibid., p. 218.
38. Idem.
39. Idem.
40. Ibid., Chapter 18.
41. Ibid., p. 306.
42. See Yuval Joseph Zacks, 'Operation Desert Storm, A Just War?', *Military Review*, January 1992, p. 20. Zacks concludes that Desert Storm met the criteria for just war. His conclusion is questioned by Ranier H. Spencer in 'A Just War Primer', *Military Review*, February 1993, p. 20, based on the targeting of infrastructure that served both Iraqi civilian and military needs. Cohen, infra n. 47, supports the targeting of such infrastructure.
43. Steven Keeva, 'Lawyers in the War Room', *The ABA Journal*, December 1991, p. 52.
44. Idem.

45. Jeffrey F. Addicott has compared the tactics of Lee and Sherman and their compliance with the laws of war (for example, Lieber Code), and then questioned General Schwarzkopf's choice of Sherman as one of his heroes. See Jeffrey F. Addicott, 'Operation Desert Storm: R. E. Lee or W. T. Sherman?', *Military Law Review*, Vol. 136, Spring 1992, pp. 115, 133.

46. Quincy Wright, *A Study of War* (Chicago: University of Chicago Press, 1942, 1971), pp. 305–307. Wright has attributed this trend to the militarization of the population and the nationalization of the war effort. 'The moral identification of the individual with the state has given the national will priority over humanitarian considerations ... consequently, the principle of military necessity has tended to be interpreted in a way to override the traditional rules of war for the protection of civilian life and property.'

47. Eliot A. Cohen, 'The Mystique of US Air Power', *Foreign Affairs* (January/February 1994), pp. 109, 123.

48. Joseph Sobran editorial, 'Modern State has Warped our Sense of Right and Wrong', *The State* (Columbia, SC), 16 February 1995, p. A19.

49. Creveld, *The Transformation of War*, supra n. 1, p. 202. Creveld has cited Mao Tse-Tung: 'Mao spoke of guerrillas as fish swimming in the sea of the surrounding population, the point of the analogy being precisely that the sea does not distinguish one part from another' (pp. 206–207).

50. Steven Metz, *America in the Third World: Strategic Alternatives and Military Implications* (US Army War College, Carlisle Barracks, PA: Strategic Studies Institute, 1994), p. 37.

51. William Raspberry quoted Havel to illustrate the importance of generic religion to contemporary culture, and its conspicuous absence from contemporary politics. William Raspberry, 'The Mystery of the Universe', *The State* (Columbia, SC), 9 July 1994, p. A8. Steve Metz has suggested that religion may be the only antidote for anarchy; see n. 50, supra.

52. In a letter written by Thomas Jefferson to Henry Fry on 17 June 1804, Jefferson left no doubt as to his love for the teachings of Jesus and his contempt for the distortion and misuse made of those teachings by preachers and politicians: 'I consider the doctrines of Jesus as delivered by himself to contain the outlines of the sublimest morality that has ever been taught; but I hold in the utmost profound detestation and execration the corruptions of it which have been invested by priestcraft and kingcraft, constituting a conspiracy of church and state against the civil and religious liberties of man.' Thomas Jefferson, *The Jefferson Bible* (New York: Clarkson N. Potter, Inc., 1964), p. 378.

53. From an introduction by Henry Wilder Foote to *The Jefferson Bible*, idem at p. 12.

54. As discussed earlier in this chapter, chivalry developed as a medieval amalgam of the Christian faith and the warrior ethic, and remains an element of military culture. Lacking conventional threats, presidents have cited it to sanctify military operations (see n. 17 supra), and individual soldiers have depended upon it to justify missions of dubious legitimacy, as in Vietnam: 'American soldiers don't go to war in the spirit of mercenaries or legionnaires; we have to think of ourselves as crusaders. It may be self-delusion, but a sense of chivalric purpose is essential to our spiritual survival when we find ourselves called upon to kill others and risk being killed', Tobias Wolff, 'After the Crusade', *Time* (24 April 1995), p. 48. The Biblical account of the Roman centurion who impressed Jesus with his faith (see Matthew 8:5–10)

27

has modern application: for the spiritual dimension of leadership, see Chapter 5, nn. 10–14, infra.

55. In a keynote address at the US Army School of the Americas at Fort Benning, GA, on 10 August 1994, entitled *The National Armed Forces as Supporters of Human Rights*, General Barry R. McCaffrey, Commander in Chief, US Southern Command (USSOUTHCOM) compared the tactics of General Sherman with those of General Robert E. Lee, to illustrate the application of the golden rule (General McCaffrey used the phrase 'treating soldiers with respect'). This is discussed under 'Operational law and human rights' in Chapter 4 (see n. 57 to Chapter 4, infra). General McCaffrey urged the Latin American officers to follow the leadership example of General Lee and treat their soldiers and civilians with respect: 'It is not always understood that soldiers treat civilians, prisoners, and other people's property as they themselves are treated. So if we treat our own soldiers with dignity under the rule of law, with some sense of compassion, then they are much more likely to act in a similar fashion toward the civilian population.' See also n. 57–60 to Chapter 4, and n. 56 to Chapter 6.

2
Might and Right in the New Millennium

We were hoping for peace – no good came of it!
For the time of healing – nothing but terror!

Jeremiah 8:15

MILITARY LEGITIMACY

Military legitimacy is a derivative of political legitimacy, which has been defined as *the willing acceptance of the right of a government to govern or of a group or agency to make and enforce decisions . . . [and] is the central concern of all parties involved in a conflict.*[1] The willing acceptance required to establish the legitimacy of military operations and activities must come from a nation's people. Public support represents that acceptance; and in a democracy public support is both a requirement and measure of military legitimacy.

Since the dissolution of the Soviet Union, public attitudes and expectations for the military have changed, and with them the requirements of military legitimacy. Civil-military relations are more important than ever in the new strategic environment. New doctrine on operations other than war recognizes the priority of legitimacy and civil-military relations in addressing threats at home and abroad.

THE NEW STRATEGIC ENVIRONMENT

Just when the world expected a new era of peace with the end of the Cold War it was reminded that peace is not the natural state of man. No sooner had the demon of Soviet communism been exorcised than a dozen lesser demons rose to take its place. A myriad of atavistic tribal, ethnic, religious and cultural (primal) conflicts have been exacerbated by overpopulation and diminishing resources. This spreading primalism has defied diplomatic or military resolution.

29

The ancient rivalries now fueling world violence were in remission during the Cold War when the competing ideologies of communism and democracy took center stage. The Cold War held local violence in check with MAD (mutual assured destruction) strategies and ever-growing nuclear arsenals. With nuclear annihilation looming over the world, it is little wonder that local enemies were temporarily forgotten, with the superpowers the larger common enemy.

Threats legitimize might, and during the Cold War a palpable and pervasive Soviet threat justified a strong US defense. The dissolution of the Soviet Union upset the strategic balance of might and right; the once clear and present threat to US survival was replaced by more violence, but no discrete threat upon which to refocus strategies. Though not a traditional threat, spreading regional violence represents a cancer that could metastasize into terminal global anarchy.

THE THREAT ENVIRONMENT: REGIONAL VIOLENCE

The new security environment is characterized by often senseless violence and human suffering that defy traditional political and military solutions. Anarchy and flagrant violations of human rights outside the US may not threaten traditional security interests, but most Americans will not tolerate them, and the media will not let us ignore them.

One common characteristic of primalism is intolerance and its corollaries – prejudice, bigotry and hate – which produce the most intractable violence. The volatile mix of religion and politics underlies conflict around the globe. From Northern Ireland to northern Africa, India, throughout Eurasia, Eastern Europe, and back to the crucible of religious conflict in the Middle East, increasingly militant religious fundamentalism – whether Christian, Jewish, Islamic, Hindu or Confucian – threatens liberty and life.

This is especially true in the case of Islamic fundamentalism. In its milder form it shares the fundamentalist belief that God has ordained a fixed set of rules and condemns divergent opinions and behavior. Virtually all militant Islamics oppose democracy and human rights.[2] The more radical elements represent an even greater threat if they ever have nuclear weapons. The damage caused by the bombing of the New York World Trade Towers was light compared to what could have been done by a tactical nuclear device.

The violent fragmentation of the former Yugoslavia has been along the medieval cultural and religious fault line separating Western and Eastern Orthodox Christianity (the Croats and Serbs) and Islam. It has been described as a clash of civilizations, and a paradigm for future conflict.[3] In nearby Germany the resurgence of neo-Nazism in response to a flood of refugees from the Balkans is an ominous indicator of renewed ethnic violence.

Russia remains the real wild card in the evolving security environment. The rise of a nationalist movement in a humiliated Russia is the most immediate threat to world peace.[4] A return to brutal autocracy would destabilize Eurasia and Eastern Europe. But at minimum, the hostility of former Soviet republics will continue toward 'Mother' Russia and those ethnic Russians living within their borders, and vice-versa. Conflict and human suffering are all but inevitable in that part of the world.

In the Pacific Rim, Japan can be expected to assume a more aggressive foreign policy role. In addition to developing new markets, Japan is likely to strengthen its defense force to provide a military capability to protect expanding security and economic interests in the region. As with Germany in Western Europe and Russia in Eurasia, the US will be obliged to respect, if not encourage, an increasing security role for Japan in the Pacific.

Africa has been unable to break a tragic cycle of civil war and natural disaster since freedom from colonial rule. Uncontrolled tribal rivalries have produced anarchy,[5] most recently at a terrible cost of lives in Rwanda. Throughout Asia, primalism has been exacerbated by the dissolution of the Soviet Union, once a stabilizing force in the region. And US humanitarian assistance to the Kurds in Operation *Provide Comfort* has strengthened their desire for a homeland, undermining political stability in Turkey as well as Iraq.

Closer to home, experiments with democracy are more promising, but problems remain. Mexico is experiencing civil violence. Panama, Nicaragua and El Salvador have made some progress toward democracy, human rights and the rule of law; but the political prognosis remains uncertain. Further South, in Venezuela, Colombia and Peru, drug wars sponsored by the US have had little success in stemming the flow of drugs into the US and have not contributed to political legitimacy in the host countries. In the Caribbean, Cuba's communist government is foundering without Soviet support, and in Haiti it is too soon to determine whether President Aristide can make democracy a practical reality.

The contemporary strategic environment presents a disturbing scenario: primal violence in the Third World, a resurgent Germany, a more aggressive Japan, a frustrated Russian Bear nursing visions of lost empire, and political oppression in the Caribbean. But external violence is only part of the threat environment. There are troubling trends within the US that could mature to threaten democracy, human rights and the rule of law.

THE THREAT WITHIN: TROUBLING TRENDS

Since the days of the Wild West the thought of internal anarchy as a threat to security has seemed far-fetched in the US. But troubling trends indicate an increasing fragmentation, frustration, intolerance, militancy and erosion of respect for the rule of law – trends that threaten the principles of democracy and human rights.

The disappearing middle class

A strong middle class has been the most significant economic stabilizing force in the US. By comparison, economic disparities in Latin America have contributed to a tradition of political violence. In the last decade, world economic trends have resulted in changes at both ends of the economic spectrum: producing more wealthy and working poor at the expense of the middle class. The increasing economic disparities do not bode well for political stability.

The changing role of government: from protecting freedom to providing entitlements

Political evolution has produced a dramatic shift in the primary purpose of government: from protecting freedom and economic opportunity to providing economic security through entitlements for social security, welfare, and health care. The pursuit of happiness is no longer enough; it has become an entitlement. Senator Bob Kerry has predicted that without some kind of reform, entitlements will constitute 70 per cent of the US budget by the year 2003. With government perceived as a pie, special interests are becoming increasingly aggressive in demanding their fair share. With an increasing demand for decreasing public resources, more conflict can be expected.

The increasing power of special interest groups

American politics began with two major political parties that polarized divergent views, a prerequisite for orderly legislative action. During the second half of the twentieth century special interest groups eclipsed political parties as power brokers, pushing the legislative process into gridlock. From the religious right to the National Association for the Advancement of Colored People (NAACP), special interest groups have fragmented and polarized political power. In the absence of some overriding issue to coalesce public opinion, such as a major threat to national security, this fragmentation of political power will continue to undermine the effectiveness and ultimately the stability of democratic political processes.

Demographics and democracy

Democracy is, by definition, rule by the majority, and that majority is a constantly shifting coalition of diverse interests. Two demographic trends are likely to test democracy in the near future: the increasing numbers of older Americans and of illegal immigrants.

Older Americans are the most formidable of all special interest groups; so much so that social security, the largest of all entitlement programs, has remained exempt from budget cuts. But something will have to give in the years ahead. With the costs of maintaining an older population going up, especially health care, and with more retired Americans and others dependent on welfare, at some point those dependent on entitlements will outnumber those paying the bill. This will be the moment of truth in the history of US democracy, and its demise if the majority of Americans continue to vote their selfish interest.

Illegal immigrants present a more immediate political crisis. The cost of providing public services to them has become so burdensome that states are suing the federal government for reimbursement. The increasing number of illegal immigrants has fed a growing xenophobia in the US that contrasts sharply with the words on the Statue of Liberty: 'Give me your huddled masses, yearning to be free . . .'

Hyphenated Americans: racial and ethnic polarization

America was once known as the melting pot. The American character was once an amalgam of many ethnic groups, all suborned to a uniquely American ideal. Today an increasingly diverse racial and ethnic population is less willing to assimilate. Ironically, the most strident advocates of separate racial identity are those African-Americans who have been there the longest. The trend toward racial and ethnic separation coupled with increasingly aggressive special interest politics promises even more polarization, with more spill-over violence likely in large cities.

Race and partisan politics

The combined effect of racial polarization and the deterioration of the two-party system has been evident in redefining partisan politics along racial lines in the South and large cities. An unlikely coalition of

33

Republicans, black caucus legislators, and sympathetic judges have promoted racially defined single-member districts to increase minority representation. The result has been to institutionalize racism as part of the political system, making racial moderates in politics an endangered species.

Intolerance in the name of political correctness

The political power of special interest groups has made public dissent unfashionable and even dangerous. With the fragmentation of America into racial and ethnic enclaves, tolerance is no longer a virtue, even at colleges and universities. The danger of the trend toward intolerance is in what it prevents rather than what it promotes: it discourages open dialogue between divergent groups, promoting polarization rather than understanding and reconciliation.

Religion and politics

Intolerance characterizes the mix of religion and politics, and for militant religious fundamentalists this mixture can be violent. While the First Amendment guarantees religious freedom there are fundamentalist religious groups in America that would establish a theocracy if given the opportunity. Louis Farrakhan's 'Nation of Islam' is one; his followers are dangerously militant but their numbers are limited. Christian fundamentalists (the religious right) are less organized but more numerous and therefore the most formidable political force and if trends continue there is likely to be more polarization of politics along religious lines.

The decline of the family

The term 'family values' has become the political rallying cry of the religious right in their promotion of traditional concepts of morality, just as the so-called pagan left advocates civil rights to legitimize non-traditional family units. The potential of these competing trends to undermine political stability cannot be underestimated. It has been felt in the military with continuing controversy over the status of homosexuals in uniform. In the cities a welfare system that subsidizes illegitimacy and fatherless families has encouraged young men to bond with urban tribes. Increasing gang violence is dramatic evidence of how the decline of the family relates to political instability.

From individual to group rights

The Bill of Rights defined individual rights; there was no recognition of group rights in the Constitution. The increasing power of special interest groups has changed the emphasis on human rights from the individual to the group. Whether based on military service (veterans' organizations), age (the American Association of Retired Persons or AARP), ethnic identity (NAACP), sex (the National Organisation of Women or NOW), religion (Christian Coalition), or sexual preference (gay rights organizations), civil rights laws have shifted from protecting individuals from unlawful discrimination to providing preferences for members of protected classes. Group rights have been legitimized by the law, undermining the concept of equal justice under law.

Decreasing respect for the rule of law

All of the above trends have contributed to decreasing public respect for the rule of law, which is the most troubling trend of all. Without the rule of law, democracy and human rights are meaningless. Public respect for the rule of law is absolutely critical to a free and democratic society. But that having been said, there is a good reason for the US public to be increasingly skeptical – if not cynical – of the rule of law.

The law is a reflection of national morality. The Constitution defined the law to safeguard constitutional democracy, protect persons and property, and resolve conflicts. Envisioned as a servant of society, the law has evolved into a burden upon society: liberty under law has been subverted by increasingly complex, expensive and ineffective laws, regulations and processes that permeate every corner of life. This subtle subversion of the law has been caused by lawmakers, lawyers, judges, and a public that has aided and abetted them in fostering unrealistic expectations of the law as a panacea for all inequities. Unless public respect for the law can be restored, the US will become increasingly vulnerable to internal disorder and ultimately anarchy.

Danger signs

While the above trends do not represent an imminent threat to internal security, they do represent potential dangers which we ignore at our peril. The convergence of any of these trends – for instance, increasing urban violence coupled with racial polarization in law enforcement agencies – would pose a real threat to internal security (law and order). As the last defense against anarchy, the military must be prepared to preserve the rule of law.

FLEXIBLE STRATEGIES FOR AN UNCERTAIN FUTURE

The new strategic environment requires new US military strategies which emphasize civil-military relations and security in environments characterized by ambiguous threats, instability and uncertainty. These strategies will be constrained by limited resources and a public intolerance for extended and indecisive military operations. Outside the US these strategies must be implemented by flexible coalitions in a dynamic community of nations more interdependent than ever before. Inside the US the military must work closely with domestic civil authorities.

There will be few pure military operations in the forseeable future: most will be inter-agency, civil-military, and coalition (combined). The need for flexible strategies and capabilities was noted by the former Chairman of the Joint Chiefs of Staff, General Colin Powell, who predicted a '... change from a focus on global war-fighting to a focus on regional contingencies', and that '... our new armed forces will be *capabilities*-oriented as well as *threat*-oriented'.[6]

Two types of strategies reminiscent of the Cold War are required for the new strategic environment: containment – defensive strategies to prevent the spread of violence, and engagement – offensive strategies to address hostile forces or conditions. The military capabilities needed to implement these strategies include combat and non-combat operations ranging from strikes and raids to nation assistance activities that promote democracy, human rights and the rule of law. These capabilities are considered operations other than war.

CAPABILITIES FOR WAR AND OPERATIONS OTHER THAN WAR

A combat capability must remain the mainstay of a strong US defense. For distant threats to US survival there should be a reconstitution combat force made up of reserves trained by an active component cadre. The readiness requirement for the reconstitution force can be as long as a year, but there must also be a quick response combat force for peacetime contingencies. This active component combat force can be at much reduced strength than during the Cold War.

For peacetime miltary operations combat forces are seldom required nor desirable. To promote peace in dangerous but strategically important areas around the world the military must complement its combat forces with those that can be closely integrated with the political and economic elements of US foreign policy in conducting activities other than war.

Future military capabilities must increasingly rely on reservists, or civilian soldiers (citizen soldier implies that active component soldiers are not citizens). Civilian soldiers must be more than combat reserves; they must also be front line forces in the civil-military activities that will predominate in the future.

In the spring-loaded Cold War environment the reserve components were truly in reserve since only active component forces could meet the readiness requirements of the expected Soviet attack – the Fulda Gap scenario. In the new strategic environment there are no such readiness constraints; but budgetary constraints will ensure that civilian soldiers have extensive roles in operations other than war.

Army doctrine in *FM 100–5* lists 13 activities as operations other than war, most of which are civil-military activities.[7] Their inclusion in *FM 100–5* represents a practical merger of special and conventional civil-military missions,[8] as well as the new category of peace operations.[9]

The following outline of activities consolidates those 13 listed in *FM 100–5* with analogous activities and missions from doctrine on special operations in low intensity conflict (LIC) and peace operations. Any doctrinal distinctions between these activities are outweighed by their common characteristics; they all share the requirements and principles of military legitimacy, which will be discussed in more detail in the following chapter.

As can be seen overleaf, all non-combat activities can be categorized as either humanitarian assistance or security assistance, as these terms are broadly defined, since peace operations involve some combination of these two basic categories of assistance. These non-combat activities can be provided at three levels: domestic, foreign national and international.

The following description of the activities of operations other than war begins with the umbrella concept of civil affairs, then domestic, foreign national, and international activities. Most activities can be considered civil affairs since their success depends upon public support, but they are not categorized as such. Special attention is given to operational law (OPLAW), since compliance with the rule of law is a prerequisite of military legitimacy.

Civil affairs

Civil affairs, or CA, is not listed as an activity in *FM 100–5*, but is designated as one of ten special operations activities.[10] CA is not limited to special operations; it refers to both the activities and the specialized military forces that interface with the civilian population in war and

ACTIVITIES OF OPERATIONS OTHER THAN WAR

I. Non-combat activities

   A. Domestic assistance to civil authorities

      1. humanitarian assistance (disaster relief)
      2. security assistance (riot control)

   B. Nation assistance

      1. Humanitarian assistance*
        a. disaster relief
        b. military civic action
        c. refugee control
        d. civil affairs*
        e. foreign internal defense*
        f. unconventional warfare*
        g. non-combatant evacuation operations or personnel recovery*

      2. Security assistance*
        a. (support to) counter-drug operations*
        b. combatting terrorism (counter-terrorism* and anti-terrorism*)
        c. support for insurgency or unconventional warfare*
        d. counter-insurgency or foreign internal defense*
        e. special reconnaissance*
        f. psychological operations*
        g. arms control
        h. show of force

   C. Peace operations (international assistance)**

      1. preventive diplomacy**
      2. peacemaking**
      3. peacekeeping**
      4. peacebuilding**

II. Combat activities

   A. Attacks and raids

   B. Direct action*

   C. Peace enforcement**

Note: Some activities other than war are also considered special operations or peace operations. Those activities listed *only* in *FM 100–5, Operations*, are not marked; those activities listed in *JCS PUB 3–05, Doctrine for Joint Special Operations*, are marked with an asterisk(*); those activities listed in *FM 100–23, Peace Operations*, are marked with two asterisks (**). *Military civic action* and *refugee control* have been added as operations other than war.

peace.[11] CA is important to military legitimacy whenever the military comes into contact with civilians, especially when public support is important to political objectives.

All four CA mission areas emphasize OPLAW compliance as a doctrinal imperative.[12] The first CA mission area supports combat operations by minimizing civilian interference with them and mobilizing human and material resources for combat support. It also supports US civil and foreign elements in humanitarian assistance and disaster relief. The priority of legitimacy in this mission area is made clear by the requirements that CA

> assist commanders, in co-ordination with the servicing staff judge advocate, in fulfilling lawful and humanitarian obligations to civil/ indigenous population, and ensuring operations are consistent with ... US law.[13]

The second CA mission area, support for special operations, includes support for insurgency (unconventional warfare or UW) and counter-insurgency (foreign internal defense or FID). Together UW and FID represent opposite sides of political warfare, euphemistically known as low-intensity conflict, or LIC. But even when bullets and not ballots are the means to gain or retain political power, ultimately that political power must be legitimized by public support:

> The struggle between the insurgent and the incumbent is over political legitimacy – who should govern and how they should govern. [Accordingly] one of the principal elements in this struggle is the effort to mobilize public support. Whoever succeeds at this will ultimately prevail.[14]

The third CA mission area is support for civil administration, the core component of nation assistance. Here CA personnel become directly involved with the political legitimacy of a host nation, usually providing specialized advice or assistance to foreign government officials based on expertise in one of the 20 CA functional areas which correspond to essential public services.[15]

Civil administration can help restore order out of chaos, or promote democracy, human rights and the rule of law in transitional democracies. CA functional specialists, such as legal and public administration specialists, can establish and oversee essential government functions; but more often they provide only advice and assistance to host-nation government officials. In wartime, civil administration in occupied countries is referred to as military government.

The fourth CA mission area is support for the domestic civil sector (military assistance to domestic civil authorities). This CA mission area involves providing civil administration (for example, public safety and public health) during domestic emergencies such as natural disasters and civil disturbances.[16] Civil-military confrontation under stressful circumstances is a constant threat in civil disturbances.

The four mission areas make CA the core of civil-military operations in peace and war. They illustrate the central role of CA in promoting the civil-military relations essential to military legitimacy. That was underscored by John O. Marsh, Jr., a former Chairman of the Reserve Forces Policy Board and Secretary of the Army, when he recalled a senior Russian military attache asking him how the US developed civilian control of the military. Marsh noted that CA 'would be *Exhibit A* of how we have done that'. He likened CA to 'a capillary action, a leavening action between combat, the transition to peace, and the reconstitution of government under the rule of law'.[17]

Secretary Marsh translated doctrinal dogma into a concise summary of the CA mission in nation assistance:

> The CA mission is to capture the roots of American Society – what this country is really all about – and to build other governments that reflect the best of the American experience.[18]

With humanitarian assistance and disaster relief missions on the increase, CA is an indispensable part of the Total Force. It has been described as 'the only part of the military force structure prepared by doctrine, training, experience, and personnel recruitment to deal with [the civilian organizations and agencies involved]'.[19]

Domestic assistance to civil authorities

Civil authorities seek emergency military assistance when either natural or man-made disasters create humanitarian or security needs that exceed their capabilities. In recent years civil authorities have requested military assistance to contain everything from chemical spills and prison riots to dangerous religious cults. Most humanitarian assistance is in the form of disaster relief or security assitance in the form of riot control; but in a major disaster a combination of both security and humanitarian assistance may be needed to restore law and order and meet critical human needs.

Domestic humanitarian assistance, or disaster relief, is a CA mission that emphasizes inter-agency and civil-military relations. The US Federal Emergency Planning Agency (FEMA) has the primary authority for disaster relief at the federal level, while individual governors have authority at the state level. Governors have immediate access to their Army National Guard (ARNG), composed primarily of combat forces. The specialized support and service support forces needed for disaster relief – CA, military police, medical, and engineers – are in the US Army Reserve (USAR), where federal law and regulations make it almost impossible for governors to use them on an emergency basis.[20]

Domestic security assistance is most often in the form of riot control; its first priority is to assist law enforcement authorities restore law and order during civil disturbances. The civil-military confrontations that characterize riot control require an emphasis on the principles of unity of effort and restraint.

Nation assistance

Outside the US, a wider range of humanitarian and security assistance activities are conducted to help friendly governments, as an extension of US foreign policy. Military nation assistance activities are only one element of foreign policy, for which the Department of State has primary responsibility. The inter-agency nature of nation assistance requires an emphasis on political objectives and unity of effort. And complex legal restrictions on military humanitarian and security assistance reflect subordination of military control to the State Department in this domain of the diplomat.

Political legitimacy is the goal of nation assistance:

> The goals of nation assistance are to promote long-term stability, to develop sound and responsive democratic institutions, to develop supportive infrastructures, to promote strong free-market economies, and to provide an environment that allows for orderly political change and economic progress.[21]

Nation assistance requires more than co-operation between military and civilian agencies; it requires their activities be a joint venture. The ambassador's country team is the model for such a joint venture; but when the military component is too large for the country team, a CA brigade can provide command and control.[22]

Humanitarian and civic assistance

Humanitarian and civil assistance (HCA) is more narrowly defined by law than the generic term humanitarian assistance. HCA refers to specific civil-military projects which are normally a part of military civic action activities conducted in underdeveloped nations. These activities are especially sensitive since they are a form of foreign aid, and represent an encroachment of the Department of Defense into the domain of the Department of State.

One of the statutory restrictions on HCA grew out of its use by the Reagan administration to support the Contra insurgency in Nicaragua. Congress responded with a military prohibition that forbade the provision of HCA (directly or indirectly) '... to any individual or group or organization engaged in military or paramilitary activity'.[23]

The military prohibition is just one of many restrictions that can create issues of military legitimacy in HCA. The real impetus for the 1986 HCA Act was not controversial assistance to the Contras but a mundane combined training exercise conducted in Honduras. A 1984 Comptroller General opinion criticized the US Southern Command (SOUTHCOM) for using military operational funds (organization and maintenance, or O & M funds, under Title 10) to construct permanent improvements that benefitted the local population as unauthorized foreign aid with no legitimate military purpose.[24]

The legislation that followed was a congressional compromise in a bureaucratic turf battle. It allowed the military limited authority to provide HCA when in conjunction with authorized military operations, but was limited to four narrow categories:

- medical, dental and veterinary care in rural areas;
- construction of rudimentary surface transportation systems;
- well-drilling and construction of basic sanitation facilities; and
- rudimentary construction and repair of public facilities.[25]

Military civic action is the predominant form of HCA. An aphorism popular during the Vietnam war reflects the need to teach self-reliance rather than risk dependency on direct aid: give a man a fish and feed him for a day; teach a man to fish and feed him for a lifetime. Military civic action is normally provided in relatively secure environments. But without law and order, humanitarian assistance cannot contribute to military legitimacy. This makes security assistance a prerequisite for military civic action and other HCA activities in violent environments.[26]

Disaster relief is emergency humanitarian assistance, and like other such activities emphasizes civil-military relations. While domestic disaster relief is provided through FEMA, overseas it is the responsibility of the Office of Foreign Disaster Assistance (OFDA). The HCA law cited above does not restrict foreign disaster relief (such as the famine relief provided in Somalia during *Restore Hope*) even though it is a form of humanitarian assistance.[27]

The CA capability to plan and implement disaster relief has proven itself overseas, while at home it is hampered by a cumbersome FEMA bureaucracy compounded by unnecessary legal restrictions on the use of federal military forces. For overseas disaster relief the Foreign Assistance Act designates the US International Development Co-operation Agency (USAID) authority to provide or co-ordinate US assistance; OFDA is the office within USAID that actually provides assistance. Under the Act, the USAID administrator may, in consultation with the Secretary of State, call upon the armed forces to assist OFDA with disaster relief.

When the US military provides overseas disaster relief under this authority, it must be reimbursed for its costs under the Economy Act. Unlike the limited military HCA authority, there is no authority for the military independently to provide overseas disaster relief. The result is that military disaster relief must be part of an overall State Department mission, much like HCA and security assistance. [28]

Refugee control is a form of humanitarian assistance that has become increasingly important in recent years: US civil-military operations have assisted and sometimes restricted Kurds in Northern Iraq, Haitians in the Caribbean, Moslems fleeing violence in Bosnia and the endless flow of refugees across Africa from famine and civil war. Issues of military legitimacy, especially those affecting human rights, have been at the forefront of these refugee control operations.

Often refugees are not welcome and their rights disregarded. In Germany the flood of refugees from Central Europe has ignited a backlash of violent xenophobia reminiscent of the Nazis. In the US there has been much debate over Haitian refugees and continuing illegal Latin American immigration. The President and US federal courts have not always agreed on the limits of refugee control and the rights of refugees; but when refugees come into the custody of the military, human rights become a mission priority.

To maintain legitimacy, peacetime standards for refugees must meet or exceed the wartime standards of the Geneva Convention regarding civilians. [29] Political issues are as important as legal issues in refugee camps. Requests for political asylum and temporary refuge can be expected and must be handled in accordance with the law and US policy to avoid embarrassing incidents. [30]

Security assistance

Security assistance is a military-to-military advisory mission often linked with humanitarian assitance as ongoing nation assistance activities. Like humanitarian assistance, it is in the domain of the State Department; its purpose is to contribute to the political legitimacy of friendly regimes by strengthening their armed forces. But, unlike humanitarian assistance, it involves military weaponry which can be misused by recipient militaries to violate human rights. Federal laws regulating security assistance focus on protecting human rights from abuse by military forces.

The primary legislation governing security assistance is the Foreign Assistance Act, which prohibits security or economic assistance to countries that 'engage in a consistent pattern of gross violations of internationally recognized human rights'. This language incorporates those rights of the Universal and American Declarations of Human Rights which have been adopted as US policy.[31]

Service regulations require US military personnel to report human rights violations through military channels. These regulations use standards from Common Article 3 of the Geneva Conventions to define a 'level of conduct that the US expects each foreign country to observe', with the following prohibited acts:

(1) Violence to life and person – in particular, murder, mutilation, cruel treatment and torture;
(2) Taking of hostages;
(3) Outrages upon personal dignity – in particular, humiliating and degrading treatment;
(4) Passing of sentences and carrying out of executions without previous judgment by a regularly constituted court, affording all the judicial guarantees that are recognized as indispensable by civilized people.[32]

Regulations further require US military personnel who observe any violations to disengage immediately and leave the area if possible, and report the incident to US authorities. They must not discuss such matters with non-US government authorities or journalists.

While security assistance is considered a military-to-military mission, its emphasis on human rights makes civil-military relations a priority concern. Security assistance is a continuing mission of the US ambassador's country team, with the security assistance officer (or group) a member of the team. For a larger force, the Security Assistance Force (SAF), provides the same integrated operational concept.[33]

Support to counter-drug operations is security assistance in the broader sense of the term, but is not subject to the Foreign Assistance Act. Counter-drug operations are more a law enforcement than a military function, and the Posse Comitatus Act prohibits federal military personnel from being used in a law enforcement capacity in the US, subject to certain limited exceptions.[34] Members of the ARNG are not subject to the restrictions of the Posse Comitatus Act while in a state status, so that they can support domestic drug interdiction without violating the law.

Overseas, US forces work in close co-operation with the Drug Enforcement Agency (DEA) to assist host nations with their counter-drug activities. These activities involve sensitive civil-military relations which emphasize political (law enforcement) objectives. But when drug lords challenge the legitimacy of a government, as in Colombia and Peru, counter-drug operations become more like counter-insurgency operations than law enforcement. Success against drug lords with private armies and public support requires a full range of military, political and economic measures to deny them legitimacy.

Like insurgents, drug lords and their terrorist squads must have a friendly sea in which to swim, to paraphrase Mao Tse-tung. Their legitimacy and ultimate survival depend upon a measure of public support or, at minimum, public apathy. Too often drug lords are afforded undeserved legitimacy and safe havens in fiefdoms where drug production is the only source of income for poor peasants. In these areas, a full range of nation assistance activities is required to isolate drug lords form their base of support – a prerequisite for their capture.

Most US counter-drug activities overseas have been advisory missions in support of local military forces, such as those in Bolivia, Colombia and Peru; but the 1989 Panama intervention (*Just Cause*) and arrest of President/General Noriega illustrated how far the US is prepared to go when a drug lord directly threatens its security interests.

Combatting terrorism provides security to a threatened population through both anti-terrorism (continuing defensive measures) and counter-terrorism (offensive operations). The effectiveness of both is measured by the level of public confidence in government to provide security from the threat of terrorism.[35]

Combatting terrorism is similar to counter-drug operations in that both are more law enforcement than military functions, unless the criminal activity threatens vital US security interests. In the US combatting terrorism is the responsibility of the FBI and overseas the responsibility of the State Department.

State-sponsored terrorism has a strategic dimension that takes it beyond law enforcement. As the hostile act of a sovereign state it justifies a US military response when it threatens American citizens or security interests. In counter-terrorism the doctrine of self-defense has sometimes been liberally interpreted to include anticipatory self-defense – essentially offensive measures – to protect the security of the US and its citizens from state-sponsored terrorist acts.

Counter-terrorism operations include hostage rescue, such as that conducted by Israeli military forces at Entebbe airport in Uganda in

1976; attacks on terrorist camps and terrorists, such as the US bombing raid on Libya in 1986 which followed a Libyan-supported terrorist attack on US servicemen in Germany; and abduction – the forcible unconsenting removal of a person by agents of one state from the territory of another state.[36] The capture of Manuel Noriega during Operation *Just Cause* for trial in the US could be considered such an abduction.

Peace operations

These include three types of activities: diplomacy (including *preventive diplomacy, peacemaking*, and *peace-building*); observing and supervising the terms of an existing peace, truce or cease-fire (*peacekeeping*); and the application of limited [offensive] military force (*peace enforcement*). Mission success in these activities is as dependent upon military legitimacy and civil-military relations as are the other activities listed above.[37]

Preventive diplomacy occurs in the earliest stages of a conflict and is in support of diplomatic efforts to mitigate the causes of violence. It can be supported by a full range of nation assistance activities or a show of force.[38]

Peacemaking occurs later in the stages of a nascent conflict; it is a step beyond preventive diplomacy but short of peacekeeping. Like other peace operations it supports diplomatic efforts to resolve the underlying conflict, and may include nation assistance activities and shows of force.[39]

Peacekeeping activities are conducted after a truce by a neutral force with the consent of all belligerents to the conflict. Their political objective is to support diplomatic efforts to reach a more permanent political settlement. Peacekeeping has been the mainstay of the UN, but in recent years questions have been raised about the extent of its peacekeeping role.

Presidential Decision Directive 25 (PDD 25), announced in May 1994, reflects the Clinton administration's view that the UN cannot make and keep world peace when hostilities exist. The directive includes seven preconditions for US military support of UN peace operations:

- the advancement of American interests;
- the availability of personnel and funds;
- the need for US participation for mission success;
- the support of Congress;
- clear objectives;
- a clear end state;
- acceptable command and control arrangements.

In addition to the above conditions for US military operations, PDD 25 provides preconditions for any US support of any UN peace operations, including economic assistance:

- a threat to international security or an urgent need for relief aid;
- a sudden interruption of democracy or a gross violation of human rights;
- clear objectives;
- consent of the parties before any forces are deployed;
- availability of sufficient money and troops;
- a mandate appropriate for the mission;
- a realistic exit strategy.

Ambassador Madeleine K. Albright, the US representative to the UN, explained the goal of PDD 25:

> The UN has not yet demonstrated the ability to respond effectively when the risk of combat is high and the level of local co-operation is low. The goal of the US policy directive is to ensure that we refrain from asking the UN to undertake missions it is not equipped to do and to help the UN succeed in missions we would like it to do.[40]

Peacekeeping is a non-combat mission conducted by combat forces. To maintain legitimacy they must emphasize restraint and resist the temptation to take the offensive, even when confronted with a hostile force. The purpose of peacekeepers is to monitor and facilitate the implementation of a truce agreement so as to achieve the political objective of a lasting peace. If peacekeepers take the offensive they lose their neutrality and compromise their legitimacy as peacekeepers.[41]

Peace enforcement involves the making rather than the keeping of peace; it is the combat complement to non-combat peacekeeping. Doctrine draws a clear distinction between peacekeeping and peace enforcement, but in practise the distinction is often fuzzy. A fragile peace can quickly deteriorate into violence, requiring a transition from defensive peacekeeping to offensive peace enforcement in order to restore the *status quo ante*.

> Peace enforcement operations include the restoration of order and stability, the protection of humanitarian assistance, the guarantee and denial of movement, the enforcement of sanctions, establishment and supervision of protected zones, and the forcible separation of belligerents.[42]

Peace-building represents the transition from hostilities to peace through nation assistance activities, and includes all the activities of nation assistance.[43]

Attacks and raids

Attacks and raids are offensive combat operations. Together with peace enforcement activities, they are the only combat activities of operations other than war. Combat operations have significantly greater risks for both military and political legitimacy than non-combat operations. When combat forces are committed to offensive operations, so is US prestige; there can be no substitute for military victory. The Weinberger Doctrine remains a relevant strategic standard for the commitment of US combat forces:

- The US should not commit forces to combat overseas unless the particular engagement or occasion is deemed vital to our national interest, or that of our allies.
- If we decide it is necessary to put combat troops into a given situation, we should do so wholeheartedly, and with the clear intention of winning.
- If we do decide to commit forces to combat overseas, we should have clearly-defined political and military objectives.
- The relationship between our objectives and the forces we have committed – their size, composition and disposition – must be continually reassessed and adjusted if necessary.
- Before the US commits forces abroad, there must be some reasonable assurance that we will have the support of the American people and their elected representatives in Congress.
- Finally, the commitment of US forces to combat should be the last resort.[44]

The Weinberger Doctrine should be read as the combat supplement to PDD 25. Together they represent US policy on strategic restraint, which will be discussed further in Chapters 3 and 6.

SUMMARY

The new strategic environment presents new challenges for US policy-makers and the military. While the pervasive Soviet threat has dissipated, in its place have sprung new, more ambiguous, but no less dangerous regional threats to world peace and security. And troubling trends hint of threats within the US. To contain the proliferation of violence and promote peace and security both at home and abroad the US must fashion innovative and flexible military strategies, capabilities and coalitions.

Peacetime military strategies and capabilities must be an integral part of US foreign policy. Operations other than war rely on civil-military activities to achieve the public support required for military legitimacy. These activities complement combat operations. They allow the military to be a positive and constructive peacetime force that can promote democracy, human rights and the rule of law.

Before new military strategies and leadership models can be developed for the challenges of the new millennium, the concept of military legitimacy and its component parts must first be understood. The next chapter describes the requirements of military legitimacy and translates them into seven principles essential for mission success in operations other than war.

NOTES

1. JCS PUB 3–07, *Doctrine for Joint Operations in Low Intensity Conflict*, the Joint Chiefs of Staff (Final Draft, January 1990), p. I–6 (hereinafter referred to as JCS PUB 3–07); also quoted in FM 100–20 (AFP 3–20), *Military Operations in Low Intensity Conflict*, Headquarters, Department of the Army and Department of the Air Force, 1 December 1989, at pp. 1–9; FM 100–5, *Operations*, Headquarters, Department of the Army, June 1993, at p. 13–4 (hereinafter referred to as FM 100–5); and FM 100–23, *Peace Operations* (draft), Headquarters, Department of the Army, 26 January 1994, at p. 1–18 (hereinafter referred to as FM 100–23). See also John B. Hunt, 'Hostilities Short of War', *Military Review* (March 1993), pp. 41, 46.
2. Judith Miller has concluded that militant Islam (Islamic fundamentalism) is a threat to democracy and human rights and should be opposed by the US. 'Despite their rhetorical commitment to democracy and pluralism, virtually all militant Islamists oppose both.' Ms. Miller cites others who see Islam by nature as 'fierce and militant', and Islamic law in opposition to the Universal Declaration of Human Rights and beyond reform. While militant Islamists may support democracy (majority rule) to attain power, they are opposed to minority rights. Judith Miller, 'The Challenge of Radical Islam', *Foreign Affairs* (Spring 1993), pp. 43, 45, 50–51.
3. Samuel P. Huntingdon, 'The Clash of Civilizations', *Foreign Affairs* (Summer 1993), p. 22.
4. Walter Laqueur sees Russian nationalism as '... one firmly believing that Russia's rightful role as a great power can only be saved by a strong authoritarian government.' Walter Laqueur, 'Russian Nationalism', *Foreign Affairs*, (Winter 1992–93), p. 103.
5. Robert D. Kaplan, 'The Coming Anarchy', *The Atlantic Monthly* (February 1994), p. 44.
6. Colin L. Powell, 'US Forces: Challenges Ahead', *Foreign Affairs* (Winter 1992–93), pp. 35, 41.
7. The activities of operations other than war are listed in FM 100–5, pp. 13–4 – 13–8.
8. Special operations missions are listed in JCS PUB 3–05, *Doctrine for Joint Special Operations* (draft), January 1994, pp. II–2 – II–15.
9. Peace operations activities are listed in FM 100–23, pp. 2–1 – 2–17.

10. The ten special operations activities are provided in 10 USC 167; see JCS PUB 3–05, supra n. 43.
11. See CA activities listed in Chapter 2, JCS PUB 3–57.
12. See JCS PUB 3–57, pp. I–1, I–7, II–3, IV–6. Army doctrine for the legal aspects of CA operations is provided in FM 27–100, *Legal Operations*, September 1991, at Chapter 11; related doctrine is found in Chapter 7 (legal operations in LIC) and Chapter 9 (legal operations in special operations).
13. See JCS PUB 3–57, at pp. II–3, 4; see also Barnes, 'Civil Affairs: Diplomat-Warriors in Contemporary Conflict', *Special Warfare* (Winter 1991), at p. 9.
14. *Joint Low Intensity Conflict Project, Final Report* (Ft Monroe, VA, TRADOC, August 1986), Chapter 4, p. 8; cited by Michael T. Klare, 'The Interventionist Impulse: US Military Doctrine for Low Intensity Warfare', *Low Intensity Conflict* (New York: Pantheon Books, 1988), pp. 75–76; also by Barnes, 'Legitimacy and the Lawyer in Low Intensity Conflict: Civil Affairs Legal Support', *The Army Lawyer* (October 1988), p. 5, n. 1.
15. Ibid. (JCS PUB 3–57) at pp. II–5, 6. The 20 Functional specialty skill areas for CA operations are divided into four groups:
    (1) *Public Administration Skill Area*, which includes public administration, public safety, public health, labor, legal, public welfare, public finance, public education and civil defense.
    (2) *Economics Skill Area*, which includes civilian supply, food and agriculture, economics and commerce, and property control.
    (3) *Public Facilities Skill Area*, which includes public works and utilities, public communications, and public transportation.
    (4) *Other Functional Specialty Skill Areas*, which include displaced persons, monuments and archives, cultural affairs, and civil information. Idem at pp. D–1 – D–3.
16. Ibid. at pp. II–6, 7.
17. See 'Civil Affairs in the Persian Gulf War', *Symposium Proceedings*, held 25–27 October 1991 at US Army John F. Kennedy Special Warfare Center and School, Fort Bragg, NC.
18. Idem.
19. Andrew S. Natsios, 'The International Humanitarian Response System', *Parameters* (Spring 1995), pp. 68, 79.
20. 42 USC 5121 *et seq.* (The Robert Stafford Disaster Relief and Emergency Assistance Act); DOD Directive 3025.1, *Use of Military Resources During Peacetime Civil Emergencies within the US*, 1980; AR 500–60, *Disaster Relief*, 1981. The complexity of the above law and regulations is discussed by Maxwell Alston in 'Military Support to Civil Authorities: New Dimensions for the 1990s', *The Officer* (October 1991), p. 28.
21. FM 100–5, p. 13–6.
22. Raymond E. Bell has proposed a CA brigade provide command and control for a regional security assistance force tailored for nation assistance. See Bell, 'To Be In Charge', *Military Review* (April 1988), p. 12.
23. 10 USC 40(c). For an overview of the law governing HCA, see draft *Operational Law Handbook* (JA 422, 1993) prepared by the Center for Law and Military Operations and the International Law Division at The Judge Advocate General's School, US Army, Charlottesville, VA, at pp. L–140, O–159, 160, and Tab V (pp. 240 *et seq.*).
24. The Comptroller General's report and resulting HCA legislation is discussed by Fran W. Walterhouse, 'Using Humanitarian Activities as a Force Multi-

plier and a Means of Promoting Stability in Developing Countries', *The Army Lawyer*, January 1993, pp. 16, 24–25. See also Barnes, 'Civic Action, Humanitarian and Civic Assistance, and Disaster Relief: Military Priorities in Low Intensity Conflict', *Special Warfare* (Fall 1989), pp. 34–35.

25. 10 USC 405. The Act provides preconditions for HCA: It requires a determination by the secretary of the military department concerned that the activities will promote the national security of both the US and the host nation, and also promote 'the specific operational readiness skills' of participating military personnel (10 USC 401(a) (1)). The Secretary of State must approve all requests for military HCA activities and later give Congress a full report on them (10 USC 401(b) (d)). Another precondition is that military HCA 'shall complement and not duplicate any other form of social or economic assistance' (10 USC 401(a) (2)). Confusion is added to complexity by three separate funding authorities that were enacted at different times and never reconciled: the first limits funding to specific appropriations for HCA as described above (see 10 USC 401(c) (1)); the second authority allows *de minimus* HCA to be funded from O & M funds for activities that have 'been commonplace on foreign exercises for decades' (10 USC 401(c) (2)); see n. 24 supra; the third authority for HCA is the Stevens Amendment, which was in effect before the 1986 Act but has never been repealed. It authorizes HCA costs from O & M funds that are 'incidental to authorized operations' (see Department of Defense Appropriation Act of 1985, Pub. L. No. 98–473, Section 8103, 98 Stat. 1837, 1942). For a discussion of the Stevens Amendment, a part of the 1985 Appropriations Act which followed the Comptroller General's Report and preceded the 1986 HCA Act (but was not repealed), see Walterhouse, n. 24 supra, p. 41. Department of Defense Directive (No. 2205.2, 6 October 1994) and Instruction (No. 2205.3, 27 January 1995) on HCA activities make the Assistant Secretary of Defense for Special Operations and Low Intensity Conflict DOD program manager for HCA.

26. For a collection of articles on military civic action, see *Winning the Peace: The Strategic Implications of Military Civic Action*, edited by John W. DePauw and George A. Luz, Strategic Studies Institute, US Army War College, Carlisle, PA, Chapter 4.

27. See Natsios, supra n. 19 at p. 78. The legal authorities for the Somalia Relief Operation cited by the DOD general counsel included the UN Charter and federal laws governing foreign disaster relief. See *Operational Law Handbook*, supra no. 23, pp. V–241–242; the federal statutes are listed on p. V–240.

28. See 22 USC 2292b and EO #12,163, Federal Register 56,678 (1979), reprinted in 22 USC 2381. For discussion see Walterhouse, supra n. 24, pp. 23–24. See also 31 USC 1535.

29. See generally, FM 27–10, Chapters 5 and 6.

30. Processing requests for political asylum and refuge are State Department responsibilities, and detailed guidance for handling such requests is provided in DOD Directive No. 2000.1, 1972, and AR 550-1, 1981, both of which are entitled *Procedures for Handling Requests for Political Asylum and Temporary Refuge*.

31. Sections 502B(b), 502B(d) (1), and 116(d) (1) of the Foreign Assistance Act of 1961 (22 USC 2304), and the Harkin Amendment to that Act, Section 116(a). There are narrow exceptions for security and economic assistance

if the President finds an improved record of human rights. See, generally, Thomas K. Emswiler, 'Security Assistance and Operations Law', *The Military Lawyer* (November 1991), p. 10.

32. See AR 12–15, para 13–3 (*Acts of Misconduct by Foreign Personnel*).
33. For the merits of a SAF in LIC, see William R. Johnson, Jr. and Eugene N. Russell, 'An Army Strategy and Structure', *Military Review* (August 1986), p. 69. For a discussion of how a CA unit might provide command and control of a SAF, see Bell, *To Be In Charge*, supra n. 22.
34. FM 100–5, 13–6. The Posse Comitatus Act is at 10 USC 1385; 18 USC 1541–1548. For other legal authorities governing civil disturbance operations, see draft *Operational Law Handbook*, n. 23 supra, at pp. S–210–211. The restrictions of the Posse Comitatus Act have been relaxed somewhat by recent legislation to allow military support of counter-drug operations in the US (National Defense Authorization Act for FY 1990, Title XI, *Drug Interdiction and Law Enforcement Support*, PL 101–456).
35. FM 100–5, p. 13–6.
36. See Abraham D. Sofaer, 'Terrorism, the Law, and the National Defense', *Special Warfare* (Fall 1989), pp. 12–25. Generally see Richard J. Erikson, *Legitimate Use of Military Force Against State-Sponsored International Terrorism* (Maxwell Air Force Base, AL: Air University Press, July 1989).
37. FM 100–23, pp. 1–2, 1–14 – 1–19. The Secretary General of the UN has described peace operations as the key elements in his 'Agenda for Peace' and his later 'Supplement to An Agenda for Peace' (see *Reports of the Secretary-General to the General Assembly and Security Council*, 17 June 1992 and 3 January 1995); see also Ruggie, n. 40 infra.
38. Ibid at p. 1–2.
39. Idem.
40. Elaine Sciolone, 'New US Peacekeeping Policy De-emphasizes Role of the UN', *New York Times*, 6 May 1994, p. A-1. John Gerard Ruggie has argued that the UN needs a more effective coercive capability for peace enforcement to deny aggression when deterrence and discussion fail; see Ruggie, 'Wandering the Void: Charting the UN's New Strategic Role', *Foreign Affairs* (November/December 1993), p. 26.
41. FM 100–23, pp. 1–3, 1–4; FM 100–5, p. 13–7. On the need for strict neutrality in peacekeeping and the contrasting requirements for peace enforcement, see William W. Allen, Antoine D. Johnson, and John T. Nelsen, II, 'Peacekeeping and Peace Enforcement Operations', *Military Review* (October 1993), pp. 53, 55, 58.
42. FM 100–23, p. 1–3; see Ruggie, n. 40 supra.
43. FM 100–23 at pp. 1–3, 1–4.
44. The six Weinberger standards were applied to *Desert Shield/Storm* by Thomas R. Dubois, 'The Weinberger Doctrine and the Liberation of Kuwait', *Parameters* (Winter 1991–92), p. 24.

3

Military Legitimacy

Virtue makes a nation great,
by sin whole races are disgraced.

Proverbs 14:34

If your enemy is hungry, give him food to eat;
if he is thirsty, give him water to drink.

Proverbs 25:21

Military legitimacy was defined at the beginning of Chapter 2. This chapter describes its components and the principles necessary to apply it to the strategies and leadership required for operations other than war.

THE REQUIREMENTS OF MILITARY LEGITIMACY

All military operations are an extension of politics by other means.[1] But the requirements of military legitimacy for operations other than war are different from those for warfighting. The civil-military issues that complicate legitimacy in peacetime are seldom an issue in wartime. To achieve mission success in operations other than war, military leaders must understand how civil-military relations and the political issues inherent in them influence the legitimacy of military operations.

Legitimacy provides the moral authority underpinning the right to act, and its requirements are derived from values, constitutions, traditions, religion, culture, the law, and public perceptions.[2] They relate to decisions made to use military force from strategic to tactical levels – from the decision of the President to deploy US forces to the decision of a soldier to pull the trigger.

Values

Values are the context of legitimacy; they are the virtues or vices that make people and their institutions what they are. Values give meaning

53

to standards by providing a frame of reference for the decisions that have an impact on issues of legitimacy. There are two categories of values that relate to military legitimacy: national and personal values.

National values are principles that have been institutionalized by common usage. The principles of democracy, human rights, and the rule of law are values enshrined by the US Constitution and have long been a common thread in the fabric of US national security strategy.[3] The relationship of these principles to military legitimacy is discussed in the next chapter.

National values are not always dominant security considerations. When vital US interests are threatened, national values have little effect on the legitimacy of combat operations; and when national survival is at stake, might makes right – at least until survival is assured. But when threats are ambiguous and political objectives predominate, the promotion of national values can be a litmus test for the legitimacy of US military might.

The Constitution is the bedrock of military legitimacy. It gives definition to democracy, human rights and the rule of law by providing a democratic framework for US government, a definition of funda-mental human rights and the mechanism for making and enforcing the rule of law.

The Constitution addresses a major issue of military legitimacy and leadership: it provides the structural framework for an authoritarian military within a democratic society. The drafters of the Constitution understood the need for a strong national defense; but they also knew that the military could be both a shield to protect freedom and demo-cracy and a sword to destroy them. To prevent a dangerous concentra-tion of military power, the Constitution provides for civilian supremacy and a separation of powers.

Civilian supremacy is provided in Article II, Section 2, which desig-nates the President as commander-in-chief; but the President's power over the military is balanced by powers conferred upon Congress in Article I, Section 8 to declare war, control military purse strings, make rules regulating the military, and make all 'necessary and proper' laws. To guard against the danger of a large professional army isolated from civilian society, the Constitution ensures a major role for civilian soldiers in the armed forces. Article I, Section 8, provides for state militias (the national guard) '... to execute the Laws of the Union, suppress Insurrections and repel Invasions'.

The Just War tradition had produced institutional values at the international level that restrain the use of force. Just cause and right intention are strategic principles, while discrimination and propor-tionality are primarily operational.[4] These principles reflect the influence of religion – specifically the Christian religion – upon the rules regulating the means and methods of warfare, and will be discussed further under the principle of restraint.

In addition to national values derived from the Constitution and the Just War tradition, there are personal values or character traits that have traditionally been associated with military leadership. The traditional values of loyalty, duty, selfless service and integrity make up the Professional Army Ethic and are recognized to be a frame of reference for ethical decision-making.[5]

Personal values influence decision-making when specific standards are inadequate, as is often the case in the ambiguous environments of peacetime operations. But because values are abstract they can produce different frames of reference for civilians and military personnel. Most civilians value individual rights and liberty above order and discipline, while the opposite is true for military personnel. All military officers have sworn allegiance to the Constitution, but their perspectives of duty and loyalty are shaped by the demands of an authoritarian military organization and its mission, where there is little tolerance for individual rights and liberty. Conflicting values can be expected; and while such conflicts have little effect in wartime, they can undermine civil-military relations and the legitimacy required for operations other than war.

Some within the military believe that collective military values should be protected from what they perceive to be deteriorating individual civilian values:

> The Army is a total institution that replaces individual values with the institution's values ... [through] powerful liminal forces which create intense comradeship and egalitarianism. The success of these processes can be attributed to the Army's unique and pervasive culture. The Army's future challenge is to monitor these liminal processes to ensure they are not accommodated to society's paradigm shift ... The more insular the military community, the better it will reinforce desirable values and minimize society's adverse influence.[6]

Recent events have illustrated conflicts of military and civilian values in sexual harassment and sexual preference. The 'Tailhook' scandal involved allegations of gross sexual harassment by Navy aviators during a convention at Las Vegas. The allegations reflected traditional 'macho' military values clearly out of sync with civilian values and the changing role of women in the military. The first casualties of the scandal were the Secretary of the Navy and top Navy lawyers who attempted in vain to 'cover up' the matter. But notwithstanding evidence that 140 officers committed acts which, at the very least, constituted 'conduct unbecoming an officer and a gentleman', no further punitive action was taken.

The role of homosexuals in the military is an even more contentious issue that will ultimately be resolved in the courts. But conflicting military and civilian values on sexual harassment and homosexuality must be reconciled to ensure civil-military relations that are conducive to military legitimacy and effectiveness:

> The services will have to review policies on acceptable conduct, on and off duty. Research on maintaining cohesion without scapegoating homosexuals and treating women as sex objects will have to be undertaken.
>
> Historically our national security and our social, legal, and constitutional practises have had to be balanced.
>
> To resist change would only make the adjustment more time-consuming and disruptive, and would itself undermine military effectiveness.
>
> The strength of our military depends ultimately upon its bonds to the people; the armed forces will be stronger the more they reflect the values and ideals of the society they serve.[7]

The challenge to military legitimacy raised by changing civilian standards of social behaviour cannot be solved by isolating the military from the civilian society it must serve. Good civil-military relations require that military and civilian personnel share the best values of our society and reject negative trends. Society benefits from such a sharing of values, and the military will not be contaminated by the experience.

Cultural standards

Conflicting military and civilian values are the result of different cultures within the US. Conflicting values are even more likely between different tribal, ethnic, racial and religious groups where culture clash is more obvious. Behaviour acceptable in Catholic Latin America may not be considered legitimate in Islamic Asia, and neither norms may conform to those of US culture. Because cultural standards are often associated with deeply held religious beliefs, their violation can produce negative emotional public reactions. Whenever public support in the area of operations is important to mission objectives, compliance with cultural norms is a requirement of military legitimacy.[8]

Cultural issues are pervasive in the increasingly violent strategic environment. Samuel Huntington has called it 'The Clash of Civilizations', and Robert D. Kaplan has referred to it as 'The Coming Anarchy'. By whatever name, increasing primal conflict has made

Cold War military strategies obsolete.[9] Huntington has argued that new strategies must place more reliance on understanding and co-operative efforts than on conventional military power. Achieving world peace and stability in the new era will require

> ... a more profound understanding of the basic religious and philo-sophical assumptions underlying other civilizations and the ways in which people in those civilizations see their interests. It will require an effort to identify elements of commonality between Western and other civilizations. For the relevant future, there will be no universal civilization, but instead a world of different civilizations, each of which will have to learn to coexist with the others.[10]

Cultural standards are often made obligatory for US military forces by incorporating them in directives, general orders and rules of engagement. But military leadership requires more than knowing the rules; it requires cultural orientation, a language capability and diplomacy in politically sensitive peacetime environments.[11]

The law

The rule of law gives meaning to the requirements of legitimacy. Unless the most important principles of legitimacy are enforced as legal standards they have no meaning. This applies to democratic processes and human rights which are incorporated in the Constitution and domestic US law, as well as to moral restraints derived from the Just War tradition which are incorporated in the law of war. All laws applicable to military operations and activities are collectively referred to as operational law, or OPLAW.

The law of war has proved to be an adequate standard of military legitimacy for wartime, but not for operations other than war. The unique characteristics of wartime operations are not present in operations other than war, one of which is the clear distinction between combatants and non-combatants. This distinction is critical to the legitimate use of force since the law of war authorizes the killing of combatants but requires humanitarian treatment of non-combatants. Because such a distinction is rarely possible in operations other than war, issues of targeting are more complex and require greater restrictions on the use of force.

Operations other than war are also subject to additional regulations by a Congress skeptical of peacetime military operations. At the strategic level it has attempted to limit the President's power as commander-in-chief through the War Powers Act, and at the operational level it has enacted a wide range of restrictions on military activities, as indicated in the previous chapter.

Law and anarchy

The law should have a new strategic priority in future conflict according to Robert Kaplan in 'The Coming Anarchy'. The law is synonymous with security, and anarchy its arch rival. Kaplan has cited West Africa as a dark harbinger of things to come, with criminal anarchy a strategic threat to world peace and stability:

> Disease, overpopulation, unprovoked crime, scarcity of resources, refugee migrations, the increasing erosion of nation-states and international borders, and the empowerment of private armies, security firms, and international drug cartels are now most tellingly demonstrated through a West African prism.[12]

Kaplan predicts that these elements have produced a new kind of war fought by people who prefer war and anarchy to peace and the rule of law – not unlike urban gangs in US cities. Kaplan cites Martin Van Creveld's *Transformation of War* as supporting Huntington on culture clash and his own views on the naivety of US strategies in Haiti and Somalia. Kaplan joins Creveld in predicting a new kind of lawless war, with religious fanaticism playing a larger role, the disintegration of nation-states and their armed forces, and urban crime developing into low-intensity conflict by coalescing along racial, religious, social, and political lines'.[13]

If Kaplan is correct in predicting the spread of primal violence and anarchy, re-establishing and preserving the rule of law will be a priority security objective in the new strategic environment. And if troubling trends in US cities are not reversed, the same seeds of anarchy could ultimately jeopardize the rule of law in the US.

Public support

Public support represents the collective public perceptions that both determine and measure military legitimacy in a democracy. In wartime the civilian public has been recognized to be the medium within which military operations occur and popular will 'the center of gravity of a nation's ability to wage war'.[14]

If popular will or public support is a nation's center of gravity in wartime, it is even more crucial to legitimacy and mission success in operations other than war. The inter-relationship between legitimacy and public support is emphasized in doctrine on special operations:

In modern conflict, legitimacy is the most crucial factor in developing and maintaining internal and international support ... Legitimacy is determined by the people of the nation and by the international community based on their collective perception of the credibility of its cause and methods. Without legitimacy and credibility, special operations will not gain the support of foreign indigenous elements, the US population or the international community.[15]

There are two publics that relate to the legitimacy of peacetime activities: the one at home and the one in the area of operations:

LIC [low-intensity conflict] is a political struggle in which ideas may be more important than arms. Therefore the US government, in co-ordination with allies and host nations, must fight for the minds of the people not only inside the host nation but also in the US and the international community. Gaining and maintaining popular con-sensus is essential.[16]

Public support for peacetime military operations varies with the perception of the threat: the greater the threat, the more likely is public support for military force. The public tends to forgive military excesses when the threat is clear, as in *Desert Storm*; but it has little tolerance for military excesses and collateral damage when the threat is more ambiguous, as was the case in Somalia.

Public support is based in large part on meeting the other require-ments of military legitimacy, but the public has its own unpredictable dynamic as well. Elected officials understand this; and in the likely absence of a clear and present threat to simplify issues of military legitimacy in peacetime, military leaders must understand the import-ance of public support to mission success.

Public support and the media

The pervasive publicity associated with the abortive US raid in Mogadishu in October 1993 was a reminder of the power of the press to shape the public opinion that determines the legitimacy of military operations.

Throughout the Cold War the relationship between the media and the military was characterized by mutual suspicion. The low point came during the Vietnam war, when a military made paranoid by a hostile media engaged in unwarranted censorship and cover-ups of military operations. The rocky relationship between the military and the media continued after Vietnam, as evidenced by the Iran-Contra affair and other incidents, primarily in Latin America, that threatened the legitimacy of US military operations.

Since *Desert Shield/Storm* the relationship between the media and military has improved. There have been few restrictions on media coverage of military operations, and until the abortive raid in Mogadishu coverage was mostly favorable, contributing to a more positive image of the military. But there have been extremes, such as in the pre-dawn darkness of December 1992 when the international press corps greeted Navy Seals on the beaches of Mogadishu with the blinding lights of network television. The farcical affair rendered night vision equipment inoperable; had there been opposition to the landing, American lives could have been lost.

This was the exception; most journalists have come to respect the danger of contemporary conflict. In both Bosnia and Somalia journalists have been among the casualties. These uncomfortable experiences have made journalists more sensitive to the need for military force to provide security for all non-combatants. When the media understands the need for military force it contributes to the public support required for military and political legitimacy – that is, so long as the standards of legitimacy are met.

The inter-relationship of legitimacy, public support, and the media make it essential to avoid 'bad press' in sensitive and unforgiving peace-time environments. This can be accomplished if military personnel understand and conform to the standards of military legitimacy and avoid conflicts between military and civilian values. The media can help maintain military legitimacy so long as the military remains a positive and constructive force promoting democracy, human rights and the rule of law.

The First Amendment to the Constitution protects a free press as the cornerstone of liberty. Despite a history of mutual suspicions between the military and the media, recent experience indicates the two can be allies. Military personnel should understand that it is short-sighted to restrict coverage of military operations for other than security reasons.

Promoting democracy, human rights and the rule of law is inextricably linked with First Amendment freedoms of expression. To the extent that the media reports the truth, it fulfills the public's right to know, an important element of military and political legitimacy.

THE PRINCIPLES OF MILITARY LEGITIMACY

Principles of operations other than war

The requirements of military legitimacy are not explicitly stated in military doctrine, but they are implicit in the principles of operations other than war listed in Chapter 13 of FM 100-5. With the addition of civil-military relations, the principles of legitimacy, objective, unity of effort, perseverance, restraint and security illustrate the characteristics of military legitimacy and provide a doctrinal context for demonstrating the relevance of legitimacy to operations other than war at strategic and operational levels. In Chapter 6 these principles are used as the frame of reference for lessons learned in legitimacy.

Evolution from low intensity conflict

Terminology for operations other than war may be new, but the principles are not. They have been adapted from the imperatives of low intensity conflict, or LIC. Conceptually, LIC doctrine remains relevant to the contemporary environment of violent peace:

> *Low intensity conflict* – (DOD) Political-military confrontation between contending states or groups below conventional war and above the routine, peaceful competition among states. It frequently involves protracted struggles of competing principles and ideologies. Low intensity conflict ranges from subversion to the use of armed force. It is waged by a combination of means employing political, economic, informational and military instruments. Low intensity conflicts are often localized, generally in the Third World, but contain regional and global security implications.[17]

The principles of operations other than war (OOTW) are derived from established LIC imperatives as shown below. This doctrinal evolution has blurred any real distinction between special operations in LIC and conventional operations other than war. And the use of the same principles in developing doctrine on peace operations indicates their universal application to the full range of military operations conducted in peacetime and conflict environments.[18]

LIC Imperatives	*Principles of OOTW*
1. legitimacy	1. legitimacy
2. primacy of the political instrument	2. objective
3. unity of effort	3. unity of effort
4. restricted use of force	4. restraint
5. perseverance	5. perseverance
6. adaptability	6. security

Legitimacy

Legitimacy is by definition the central concern of all parties directly involved in a conflict, making it the first and foremost principle of operations other than war.[19] Military legitimacy is not a doctrinal term; it was coined to relate the concept of legitimacy to military operations and leadership. It reflects the central role of civil-military relations to mission success in peacetime.[20]

Military doctrine begins by warning military leaders that legitimacy cannot be imposed by force of arms, and that short-cut attempts to solve problems related to legitimacy may compromise strategic objectives:

> Legitimacy derives from the perception that [military] authority is genuine, effective and uses proper agencies for reasonable purposes. No group or force can decree [or force] legitimacy for itself [or others], but it can create and sustain legitimacy by its actions.[21]

LIC doctrine emphasizes legitimacy as an essential ingredient of leadership, and confirms the importance of public support to strategic objectives:

> In order to accomplish their larger objectives in LIC, military leaders must consider the effect of all their actions on public opinion. The legitimacy of the actions of an armed force, or even individual members of the force can have far-reaching effects on the legitimacy of the political system that the force supports. The leader must ensure that his [or her] troops understand that a tactically successful operation can also be strategically counterproductive because of the way in which they executed it and how the people perceived its execution.[22]

The remaining principles are inextricably bound with the concept of military legitimacy. As illustrated by the following doctrine, these principles relate the requirements of legitimacy to operational and strategic issues.

Objective

Objective is the second principle, but it should be read as political objectives based on LIC imperatives of primacy of the political instrument and political dominance.[23] It is the predominance of political over military objectives in peacetime that makes legitimacy the central concern of all parties in a conflict. Compliance with all requirements of military legitimacy is necessary to achieve political objectives, with a special emphasis on the law, cultural norms, national values and public support.

Objective also refers to the 'end state' of a military mission: the required conditions for attaining strategic objectives or for passing the mission on to another instrument of national power. In operations other than war as well as special operations in LIC and peace operations, the end state relates the mission to US or international political objectives. Military operations will always complement diplomatic, economic or humanitarian efforts.[24]

Joint doctrine on LIC emphasizes the importance of political objectives and cultural standards to effective leadership, even to the point of recommending that leaders not feel bound by the constraints of traditional doctrine to achieve legitimacy:

> In LIC operations political objectives drive military decisions at every level from the strategic to the tactical. Commanders and their staff officers must understand the specific political objectives and the impact of military operations on them. They must adopt courses of action which legally support those objectives even if the courses of action appear to be outside what traditional military doctrine encompasses. Such an approach demands an understanding of the host nation culture, customs, and policies. A principal source of advice in this area is the civil affairs staff officer and the supporting civil affairs unit commander.[25]

The principle of objective requires that the end state of peacetime operations be clearly defined at the strategic level to prevent 'mission creep', the dangerous tendency for commanders to shift missions – usually using more force than authorized – when it appears expedient. Such mission clarity is more difficult in peacetime than wartime.

Unlike wartime military objectives that can be defined in clear and quantitative terms (for example, taking and holding terrain), the end states of most peacetime operations have political objectives that are more qualitative and ambiguous. This ambiguity results in a tendency for military commanders to substitute inappropriate military objectives and criteria for success, which in turn can lead to the excessive use of force and loss of legitimacy. The body count in Vietnam was such an inappropriate measure of success, and the same tendency to substitute military criteria for political objectives was evident in US combat operations in Somalia.

The need for clarity in defining political objectives in military terms is closely related to the principles of unity of effort and restraint discussed below:

A clearly defined and attainable objective – with a precise under-standing of what constitutes success – is critical when the US is involved in peace operations. Military commanders should also understand what specific conditions could result in mission failure as well as those that yield success. Commanders must understand the strategic aims, set appropriate objectives, and ensure that these aims and objectives contribute to unity of effort with other agencies.[26]

Unity of effort

This is the third principle of operations other than war. It has an organizational orientation, but is closely related to political objectives and civil-military relations. Unity of effort is analogous with unity of command, a principle of combat operations; but operations other than war involve a wide variety of military and civilian personnel, which makes traditional concepts of unity of command impractical. Still, mission success depends upon a co-ordinated effort, requiring military leaders who can combine military proficiency with the finesse of a diplomat:

> Commanders may answer to a civilian chief, such as an ambassador, or may themselves employ the resources of a civilian agency. Command relationships may often be only loosely defined, causing commanders to seek an atmosphere of co-operation rather than command authority to achieve objectives by unity of effort. Military commanders consider how their actions contribute to initiatives that are also political, economic, and psychological in nature.[27]

The international dimension of many operations other than war, especially peace operations, makes unity of effort even more complex:

> Whenever possible, commanders should seek to establish a com-mand structure that takes account of, and provides coherence to, all activities in the area. As well as military operations this command structure should include the political, civil, administrative, legal, and humanitarian activities involved in the peace operations. With-out such a command structure military commanders need to consider how their actions contribute to initiatives that are also diplomatic, economic, and informational in nature. This will necessitate extensive liaison with all the involved parties as well as reliable communica-tions. Because peace operations will often be conducted at the small-unit level it is important that all levels understand the military-civilian relationship to avoid unnecessary and counterproductive friction.[28]

The demanding qualities of leadership required to achieve unity of effort in operations other than war require bridging the formidable gap between military and diplomatic matters as well as overcoming barriers of culture and language. These exceptional requirements of leadership will be discussed in Chapter 5.

Perseverance

Perseverance, or patience, is the fourth principle of operations other than war. Political objectives do not often lend themselves to a quick-fix, but US law and political expediency have traditionally favored short and decisive applications of combat force in peacetime. While perseverance may be an ideal, short-term operations are more likely to be the norm in the future. The US public has traditionally been impatient with peacetime military operations, especially when they involve American casualties.

This impatience is reflected in the law: The *War Powers Act* (50 USC 1541–1548) requires the president to consult with Congress before committing US forces '... where imminent involvement in hostilities is clearly indicated by the circumstances ...' and limits US military involvement in such situations to 60 days without the approval of Congress. The US strikes into Grenada (*Urgent Fury*) and Panama (*Just Cause*) reflect the bias of presidents and Congress towards quick and dirty combat solutions to peacetime security issues rather than more protracted and controversial alternatives, such as nation assistance, that rely on non-combat activities.[29]

External military force cannot create the political legitimacy required for lasting peace: that legitimacy ultimately depends upon broad-based indigenous public support, a condition lacking in contemporary conflicts based on intractable cultural (religious and ethnic) differences. Helping indigenous forces establish law and order out of chaos and then mobilizing the public support required for stable government depends upon long-range political objectives and the perseverance to achieve them. Army doctrine cautions commanders to be patient in such situations, avoiding combat if it threatens long-term strategic objectives:

> Commanders must assess quick contingency response options against their contribution to long-term, strategic objectives. This does not preclude decisive military action but does require careful, informed analysis to choose the right time and place for such action.[30]

Commanders are warned that short-sighted quick fixes can threaten strategic aims:

If committed forces solve an immediate problem within a nation or region but detract from the legitimacy of the government in so doing, they have acted detrimentally against long-term, strategic aims.[31]

Perseverance is especially important to legitimacy in peace operations:

Commanders balance their desire to attain objectives quickly with a sensitivity for the long-term strategic aims and the restraints placed on [peace] operations. This principle requires patience and the willingness to amend traditional measures of success and victory with the new ones that gauge social and political progress, preventions of humanitarian catastrophe, and considerations of post-conflict peace-building measures.[32]

Restraint

The fifth principle, restraint, is the most important component of military legitimacy relating to the use of force. Excessive force can cause collateral damage which undermines the public support required for political objectives.

Restraint at both strategic and operational levels is based on principles derived from the Just War tradition: discrimination and proportionality.

Discrimination relates to the choice of targets at operational and tactical levels. The same principles considered in determining whether lethal force can legitimately be used against a specific target to achieve military objectives can be used at the strategic level to determine *just cause* – whether military force is a legitimate means to achieve national political objectives.

Proportionality requires that military force be proportional to military objectives at operational and tactical levels; it is closely related to *right intention*, which applies proportionality to balance the military means and methods used to achieve political objectives at the strategic level. Right intention requires the least force necessary to achieve strategic objectives. Both right intention and proportionality require more restraint on the use of lethal force in peacetime than in wartime.

Standards of restraint are incorporated in rules of engagement (ROE) which are tailored to each operation. Peacetime ROE are much more restrictive than wartime ROE because of the primacy of political objectives and the need for public support to achieve them. ROE reflect offensive military force to be the norm in wartime with defensive force the norm in peacetime.[33]

In operations other than war, these ROE will be more restrictive, detailed, and sensitive to political concerns than in war. Moreover, these rules may change frequently. Restraints on weaponry, tactics, and levels of violence characterize the environment. The use of excessive force could adversely affect efforts to gain legitimacy and impede the attainment of both short-term and long-term goals.[34]

The restricted use of force has long been an LIC imperative which underscores the importance of the law and ROE as requirements of military legitimacy:

The nature of the LIC environment imposes greater limits on the use of military power than is usually the case with conventional warfare. This is reflected in the legal restrictions and the operational and social restraints usually encountered in LIC. Military operations in the LIC environment may be highly visible and politically sensitive. They require particular attention to international, US, and host nation law including multinational and bilateral agreements and Congressional authorizations and appropriations. ...

Excessive violence can adversely affect efforts to gain or maintain legitimacy and impede the attainment of both short-term and long-term goals.[35]

In peace operations the principle of restraint is especially important to legitimacy:

Restraints on weaponry, tactics, and levels of violence characterize the environment. The use of excessive force will adversely affect efforts to gain or maintain legitimacy and impede the attainment of both short- and long-term goals. The ROE, and reasons for them, need to be understood and regularly practised by all soldiers since a single thoughtless act could have critical political consequences.[36]

In peace operations there must be a clear distinction between the defensive and offensive use of force. Self-defense is the norm, but even when offensive force is authorized the principle of proportionality requires that leaders exercise restraint:

In peacekeeping operations, force will be used only in self-defense. In peace enforcement operations the use of [offensive] armed force may prove necessary, but presents many dangers. Armed action will almost certainly prejudice the acceptability [e.g., legitimacy] of the troops carrying it out and could have far-reaching international consequences. Since force invariably attracts a response in kind, its use may also escalate tension and violence in the local area and embroil peace operation troops in a harmful long-term conflict that is irrelevant to their aims. For that reason the use of force will always be a last resort and should only be used when all other means of persuasion are exhausted. Decisions made by commanders concerning the use of force are likely to be critical to the nature and conduct of operations.

In all cases the application of force will be proportional to the threat and restrained. In peace operations every soldier must be aware that the goal is to produce conditions which are conducive to peace and not to destroy an enemy.[37]

At the strategic level, decisions involving the commitment of US forces require the exercise of restraint. These strategic decisions involve complex ethical, moral and political issues which require the application of the requirements of legitimacy, particularly the law, national values, and the principles of just cause, discrimination and proportionality:

LIC more than war will often present the US and its armed forces with difficult ethical and moral decisions. The type of aggression encountered in LIC is not as blatant as that in war. Subversion, sabotage, assassination, and guerrilla operations encountered in another country may pose a threat to US interests, but the threat to national survival may be neither imminent nor obvious. The US response to this threat must be consistent with US and international law and US national values. The response of the US to these threats may be controversial because there may be legitimate grievances that provoke them. Nonetheless, the decision to stand aside is as profound in its effect as the decision to become involved.

The decision to act ... is essentially a political one. International law and custom presume that an incumbent government is legitimate and legally constituted. *A policy of involvement by an outside power must demonstrate its legitimacy. The basis for the international use of force is self-defense or the defense of others.* [emphasis added][38]

The Weinberger Doctrine and PDD 25 reflect a US policy of strategic restraint. They have adapted the principles of just cause (strategic discrimination) and right intention (strategic proportionality) to make combat the military measure of last resort in peacetime. Just cause for the commitment of combat forces is dependent upon a threat to US survival or vital security interests. Right intention requires that the size, composition and disposition of military force be proportional to the threat, and that US combat forces are the military means of last resort. In addition to restraint, the Weinberger Doctrine emphasizes the need for clear political and military objectives and the public support and perseverance required to achieve them.

General Colin Powell, former Chairman of the Joint Chiefs of Staff, has attributed US successes in peacetime engagement to understanding the primacy of political objectives and co-ordinating military operations with diplomatic and economic efforts to achieve those objectives:

> The reason for our success is that in every instance we have carefully matched the use of military force to our political objectives ... When force is used deftly – in smooth co-ordination with diplomatic and economic policy – bullets may never have to fly. Pulling triggers should always be toward the end of the plan.[39]

Security

Security, the sixth principle of operations other than war, complements the principle of restraint. While lethal force must be restrained to achieve political objectives in peacetime operations, that restraint must be balanced with the need for security, or self-defense:

> Regardless of their mission, commanders must protect their forces at all times. The intrinsic right of self-defense always applies.[40]

Commanders may be responsible for protecting more than their own commands; they may also be responsible for the security of civilians in their area of operations. If so, the requirements of civil law and order make diplomacy and good civil-military relations prerequisites for legitimacy:

> Security ... derives from more than physical protective measures. A force's security will be significantly enhanced by the force's perceived legitimacy and impartiality, the mutual respect built between the force and the other parties involved in the peace operation, and the force's credibility in the international arena, which derives from an effective public affairs, psychological operations and civil affairs program.
>
> In a peace operation context, force projection may extend beyond the commander's forces. Security may have to be provided while performing civic and humanitarian projects and may be extended to civil agencies and non-governmental agencies.[41]

As it relates to legitimacy and the public support required for political objectives, security applies more to civilians than military forces. Providing security for persons and property is the first priority of political legitimacy. In the absence of effective civil law enforcement the military has a moral if not a legal obligation to provide security as an integral part of all operations other than war.

> Security is a central facet of the war of moral legitimacy ... [It] must be linked to village-level self-development programs supported by civic action and backed by regular military forces.[42]

If the predictions of Robert Kaplan in 'The Coming Anarchy' prove to be correct, primal violence and the disintegration of traditional military forces could make security the predominant requirement of military legitimacy in future operations other than war – a contrast with the predominance of restraint in the past.

Civil-military relations

Civil-military relations are not included in FM 100-5, but should be the seventh principle of operations other than war. If there is one dominant characteristic of military operations other than war that distinguishes them from war-fighting, it is that they focus on civilians rather than enemy combatants. The inter-relationship between the military and civilians can make the difference between military victory and political defeat in peacetime military operations.

In wartime, civil-military relations are secondary to defeating the enemy with overwhelming force. But in peacetime, when public support for political objectives both at home and in the area of operations is more important than defeating an ambiguous enemy, civil-military relations become the primary focus of legitimacy.

The primacy of civil-military relations in peacetime explains why most operations other than war are civil-military operations. They reflect all the principles discussed above: the public support and perseverance required to achieve political objectives, unity of effort (especially with civilian leaders), restraints on the use of force to avoid collateral damage and security for the civilian populace.

Civil affairs, or CA, refers to both civil-military operations and the forces that conduct them.[43] As mentioned in the previous chapter, CA provides the interface between military forces and civilians that is critical to legitimacy in operations other than war. CA doctrine emphasizes civil-military relations as a command responsibility and includes those activities described in Chapter 2 that project military leaders into the unfamiliar domain of civilian politics:

> Civil Affairs is a responsibility of command. CA involves all activities associated with the relationship between military forces and civil

authorities and population in a friendly or occupied country or area where military forces are stationed or employed.[44]

More than any other military discipline, CA requires compliance with the law as a mission objective, specifically,

> to assist command compliance with OPLAW requirements by providing those resources necessary to meet essential civil requirements, avoiding property and other damages to usable resources, and minimizing loss of life and suffering in so far as military circumstances permit.[45]

The requirements and principles of legitimacy in operations other than war underscore the importance of CA as both a concept and capability needed for mission success in the new strategic environment.

SUMMARY

The requirements of military legitimacy reflect the *right* that must take precedence over *might* to ensure mission success in operations other than war. Values, cultural and legal standards, and public support provide standards and a context for decision-making. These requirements are different in times of war and peace due to the predominance of civil-military issues and political objectives in peacetime. Determining what might is right in ambiguous and unforgiving peacetime environments requires an understanding of the concept of military legitimacy and its requirements.

The doctrinal principles of operations other than war, with the important addition of civil-military relations, illustrate the relevance of the requirements of military legitimacy to mission success. These principles of military legitimacy emphasize the predominance of political objectives, and the need for unity of effort (especially interagency and civil-military activities) and perseverance to mobilize the public support required to achieve those objectives. Balancing restraint in the use of force with the requirements of security to achieve military and political objectives without sacrificing public support is the greatest challenge of military legitimacy. When public support is essential to mission success civil-military relations are the measure of victory or defeat; and civil affairs represents both civil-military operations and the forces that conduct them.

The national values of democracy, human rights and the rule of law are at the core of military legitimacy. They are the virtues that gave US national strategy the moral highground over communism during the

Cold War; but their application in the new strategic environment raises unanswered questions that will challenge civilian and military leaders in the new millennium.

NOTES

1. This principle is usually attributed to Carl von Clausewitz. See Clausewitz, *On War*, edited and translated by Michael Howard and Peter Paret (Princeton, N.J.: Princeton University Press, 1984), p. 84. In spite of his recognition of the political objectives of warfare, General Clausewitz never considered the viability of operations other than war. He was convinced that once the military became an instrument of national policy, the only option was to apply unlimited force. While Clausewitz opened the door to an understanding of the political nature of warfare, his failure to appreciate the limited use of military force renders his classic strategy inapplicable to operations other than war. See B. H. Liddel Hart, 'National Object and Military Aim', *Strategy*, 2d rev. ed. (New York: Frederick A. Praeger, 1967), Chapter 21; also Sam C. Sarkesian, 'Organizational Strategy and Low Intensity Conflicts', *Special Operations in US Strategy*, edited by Frank R. Barnett, B. Hugh Tovar, and Richard H. Shultz (New York: National Defense University Press, National Strategic Information Center, Inc., 1984), p. 274.
2. Barry Crane et al., 'Between Peace and War: Comprehending Low Intensity Conflict', *Special Warfare* (Summer 1989), p. 5.
3. Former President Bush emphasized the importance of national values to US security policy in the preface of the *National Security Strategy of the United States* (The White House, August 1991): '[O]ur values are the link between our past and our future, between our domestic life and our foreign policy, between our power and our purpose. It is our deepest belief that all nations and peoples seek political and economic freedom; that governments must rest their rightful authority on the consent of the governed, and must live in peace with their neighbors.' President Clinton has confirmed the national values of democracy, human rights, and the rule of law as principles of US security policy. See Stephen John Stedman in 'The New Interventionists', *Foreign Affairs* (America Around the World 1992/1993), p. 1.
4. See FM 27–10, *The Law of Land Warfare*, Department of the Army (1956), para 3a (military necessity); para 34 (unnecessary suffering); and para 41 (proportionality); for the application of the principles of discrimination and proportionality to peacetime military operations, see William V. O'Brien, *Special Operations in the 1980s: American Legal, Political, and Cultural Constraints, Special Operations in US Strategy* (National Defense University Press, 1984), pp. 53, 69–73. See also, *Student Text on Military Law and Justice*, Required readings in Military Science IV, Department of Law, US Military Academy, West Point, NY (June 1992), Chapter 5.
5. See FM 100–1, *The Army*, Headquarters, Department of the Army (May 1986), Chapter 4; and FM 22–100, *Military Leadership* (co-ordinating draft), Department of the Army, June 1988), Chapter 4.
6. Robert L. Maginnis, 'A Chasm of Values', *Military Review* (February 1993), pp. 2–11. Maginnis defines liminality as a technical pyschological term for rite of passage. After arguing that 'societal trends indicate a fundamental

change in national values [that are] significantly different than the Army's values', 'Maginnis issues a call to arms to isolate the military from changing civilian values: 'The Army must preserve its integrity as an institution by resisting any tendency to accommodate these changed values.' The danger of such a paranoid view of civilian values to military legitimacy and leadership is discussed in Chapter 5.

7. Richard H. Kohn, 'Women in Combat, Homosexuals in Uniform: The Challenge of Military Leadership', *Parameters* (Spring 1993), pp. 2–4.

8. On the importance of local public support to mission success see FM 100–25, *Doctrine for Army Special Operations Forces* (Final Draft, October 1990), pp. 2–29, 30. The new FM 100–5 reaffirms the need for military forces to sustain the legitimacy of the operation and the host government, and the danger that military success that detracts from political legitimacy may undermine long-term strategic aims. FM 100–5, p. 13–4.

9. Samuel P. Huntington, 'The Clash of Civilizations', *Foreign Affairs* (Summer 1993), p. 22; Robert D. Kaplan, 'The Coming Anarchy', *The Atlantic Monthly* (February 1994), p. 44.

10. Huntington, 'The Clash of Civilizations', supra n. 9, p. 49.

11. On cultural issues in LIC, James K. Bruton and Wayne D. Zajac, 'Cultural Interaction: The Forgotten Dimension of Low Intensity Conflict', *Special Warfare* (April 1988), p. 29.

12. Kaplan, supra, n. 9, p. 46.

13. Ibid., pp. 72–74.

14. A. J. Bacevich has emphasized the role of 'the people' in contemporary warfare in 'New Rules: Modern War and Military Professionalism', *Parameters* (December 1990), pp. 12, 19. On the role of public support in the Vietnam War, see Harry G. Summers, Jr., *On Strategy: The Vietnam War in Context*, Strategic Studies Institute, US Army War College, Carlisle Barracks, PA, 1989, p. 4.

15. FM 100–25, *Doctrine for Army Special Operations Forces* (Final Draft, October 1990), pp. 2–29, 30. The new FM 100–5 reaffirms the need for military forces to sustain the legitimacy of the operation and the host government, and the danger that military success which detracts from political legitimacy may undermine long-term strategic aims. FM 100–5, p. 13–4.

16. JCS PUB 3–07, see n. 1 to Chapter 2, p. I–19.

17. JCS PUB 1–02, *Department of Defense Dictionary of Military and Associated Terms*, 1 December 1989. For application of LIC to operations other than war, see Horace L. Hunter, Jr., 'Ethnic Conflict and Operations Other Than War', *Military Review* (November 1993), p. 18; James R. Locher, II, 'Low Intensity Conflict: Challenges for the 1990s', *Defense 91* (July/August 1991), p. 19; also, Barnes, 'The Politics of LIC', *Military Review* (February 1988), p. 2.

18. The doctrinal principles and imperatives are set forth in the references in n. 1 to Chapter 2.

19. See JCS PUB 3–07, p. I–26.

20. See, generally, Barnes, 'Military Legitimacy and the Diplomat Warrior', *Small Wars and Insurgencies* (Spring–Summer 1993), p. 1.

21. FM 100–5, p. 13–4; JCS PUB 3–07, p. I–26; FM 100–20, p. I–9.

22. FM 100–20, p. I–85.

23. As to primacy of the military instrument, see JCS PUB 3–07, p. I–25; as to political dominance, see FM 100–20, p. I–8.

24. FM 100–23, p. 1–8.
25. JCS PUB 3–07, pp. I–25. On cultural issues in LIC, James K. Bruton and Wayne D. Zajac, 'Cultural Interaction: The Forgotten Dimension of Low Intensity Conflict', *Special Warfare* (April 1988), p. 29. In his list of principles governing hostilities short of war, John B. Hunt did not use the term 'objective', but instead used the 'primacy of the political instrument'; they are interchangeable concepts. See Hunt, 'Hostilities Short of War', supra n. 1 to Chapter 2, at pp. 44–45.
26. FM 100–23, p. 1–14.
27. FM 100–5, p. 13–4.
28. FM 100–23, p. 1–15.
29. See, generally, Barnes, 'Military Legitimacy and the Diplomat Warrior', supra, n. 20, pp. 14–15.
30. FM 100–5, p. 13–4. See also Horace L. Hunter, Jr., 'Ethnic Conflict and Operations Other Than War', *Military Review*, supra n. 17, pp. 18, 20; also Hunt, 'Hostilities Short of War', supra, n. 1 to Chapter 2, p. 46; and Barnes, 'The Politics of LIC', supra, n. 17, pp. 4–8.
31. FM 100–5, p. 13–4.
32. FM 100–23, p. 1–18.
33. Rules of engagement (ROE) are directives issued by the competent military authority which delineate the circumstances and limitations under which US forces will initiate and/or continue combat engagement with other forces encountered. *Department of Defense Dictionary of Military and Associated Terms*, Joint Pub 1–02, 1 December 1989. For a comprehensive discussion of ROE, see *OPLAW Handbook*, at n. 23 of Chapter 3, pp. H–92 *et seq.*, and Mark S. Martins, 'Rules of Engagement for Land Forces', *Military Law Review*, Vol. 143 (Winter 1994), p. 4. On human rights generally, see *OPLAW Handbook* at Tab W (pp. W–247 *et seq.*). The ROE of *Desert Shield* (peacetime) and *Desert Storm* (wartime) illustrate the contrasting requirements of restraint in peacetime; a summary of each was distributed to soldiers in the Gulf, and reprinted in *Student Text on Military Law and Justice, Required Readings in Military Science IV*, Department of Law, US Military Academy at West Point (1992), pp. 3–4, 3–5. For a discussion of operational law and ROE in *Desert Storm*, see Steven Keeva, 'Lawyers in the War Room', *The ABA Journal* (December 1991), p. 52. By way of contrast, see a discussion of the legal issues and ROE required in Somalia in F. M. Lorenz, 'Law and Anarchy in Somalia', *Parameters* (Winter 1993–94), p. 27, and Jonathan T. Dwarken, 'Rules of Engagement: Lessons from Restore Hope', *Military Review* (September 1994), p. 26. For the application of the principles of discrimination and proportionality to peacetime military operations, see William V. O'Brien, 'Special Operations in the 1980s: American Legal, Political, and Cultural Constraints', *Special Operations in US Strategy* (National Defense University Press, 1984), pp. 53, 58; also Barnes, 'Military Legitimacy and the Diplomat Warrior', supra, n. 20, pp. 11–16.
34. FM 100–5, p. 13–4.
35. JCS PUB 3–07, pp I–24; I–27.
36. FM 100–23, p. 1–16.
37. FM 100–23, pp. 1–16, 1–17; see Dwarken article, n. 33, supra.
38. FM 100–20, pp. 1–13, 1–14.
39. The six Weinberger standards were applied to *Desert Shield/Storm* by Thomas R. Dubois, 'The Weinberger Doctrine and the Liberation of Kuwait',

Parameters (Winter 1991–92), p. 24.

40. FM 100–5, p. 13–4.
41. FM 100–23, p. 1–16.
42. John T. Fischel and Edmund S. Cowan, 'Civil-Military Operations and the War for Moral Legitimacy in Latin America', *Military Review* (January 1988), p. 41. The authors use the term civil-military operations which has essentially the same meaning as civil affairs in this context.
43. Civil affairs (CA) and civil-military operations (CMO) are closely related terms that are used interchangeably in this context. CA involves all activities involving the interface between military and civilian personnel; it is defined by law as a special operations activity. CMO is a generic term referring to the use of military forces to perform traditionally non-military activities. See JCS PUB 3–57, *Doctrine for Joint Civil Affairs*, Final Draft, November 1990, Chapter 1, paras 1 and 3; discussed in Barnes, 'Military Legitimacy and the Diplomat Warrior', supra n. 20, p. 7. See also FM 100–5, pp. 2–21.
44. JCS PUB 3–57, *Doctrine for Joint Civil Affairs* (final draft), November 1990, pp. I–1, I–2 (hereinafter JCS PUB 3–57). For an overview of CA activities as they relate to legitimacy in a peacetime environment, see Barnes, 'Civil Affairs: Diplomat Warriors in Contemporary Conflict', *Special Warfare* (Winter 1991), pp. 4, 6–9.
45. JCS PUB 3–57, p. I–7.

4
Democracy, Human Rights and the Rule of Law

Do not say I will treat him as he has treated me;

Proverbs 24:29

Always treat others as you would like them to treat you;
that is the meaning of the law ...

Matthew 7:12

THE EVOLUTION OF NATIONAL VALUES IN DIPLOMACY

Democracy, human rights and the rule of law have been US values since the birth of the nation and are enshrined in the Constitution. But they have not always been the guiding light of US national security policy. Henry Kissinger has traced the evolution of US diplomacy and its relationship to these values, contrasting the moral and sometimes theological quality of American foreign policy with the more pragmatic balance of power diplomacy (*realpolitik*) which has been prevalent among European nations since the Treaty of Westphalia in 1648.[1]

Cardinal de Richelieu, as first minister of France, introduced *realpolitik* to Europe with the concept of *raison d'état*, and he acknowledged it to be the diplomatic equivalent of the maxim that might makes right:

> In matters of state, he who has the power often has the right, and he who is weak can only with difficulty keep from being wrong in the opinion of the majority of the world.[2]

Kissinger noted little moral difference between Richelieu's *raison d'état* and the prevailing concept of Just War which sanctified wars of that period based on their intent. The will of warfare was all-important to its legitimacy: those who participated in wars that were intended to kill the guilty could not be held accountable for harming the innocent.[3] It might be said that religion had given war a bad name, and Richelieu

77

at least restored reason to warfare that had been as irrational as it had been inhumane.

Early American presidents had nothing but contempt for Richelieu and his *realpolitik* successors who promoted the view that the ends of the state justified the means:

> America ascribed the frequency of European wars to the prevalence of governmental institutions which denied the values of freedom and human dignity.[4]

Thomas Jefferson rejected the European idea that morality was for people and not nation-states, and advocated the promotion of democracy, human rights and the rule of law. But Jefferson and other nineteenth-century presidents felt that these values could best be promoted by example, not by military force:

> In the words of Thomas Jefferson, a 'just and solid republican government' in America would be 'a standing monument and example' for all the peoples of the world.[5]

Following the War Between the States and an era of US imperialism, Woodrow Wilson put democratic ideals at the heart of US foreign policy, and following the First World War these ideals were force-fed to a war-weary Europe through the Treaty of Versailles. Wilson's unbounded idealism went beyond promoting equal rights for people; he promoted equal rights among nation-states, to be institutionalized in a world organization that would provide collective security through the enforcement of international law. The League of Nations was the precursor of the United Nations, but its failure to prevent the outbreak of the Second World War testified to the limits of idealism in foreign policy.

The Second World War illustrated the irrelevance of national values in total war; there is no substitute for victory when national survival is threatened. But shortly after ordering the *Enola Gay* to vaporize Hiroshima, President Truman revived national moral values as a driving force behind US foreign policy:

> American foreign policy, as a reflection of the nation's moral values, was 'based firmly on fundamental principles of righteousness and justice', and on refusing to 'compromise with evil'. Invoking America's traditional equation of private with public morality, Truman promised that 'we shall not relent in our efforts to bring the Golden Rule into the international affairs of the world.'[6]

The Korean War was the first real test of US military resolve in the Cold War. As the primary justification for entering the conflict,

President Truman cited the need to enforce the rule of law to prevent 'a return to the rule of force in international politics'. Kissinger interpreted this as a reaffirmation of the continuing commitment of America to its national values when deploying its military forces:

> That America defends principle, not interests, law, and not power, has been a nearly sacrosanct tenet of America's rationale in committing its military forces, from the time of the two world wars through the escalation of its involvement in Vietnam in 1965 and the Gulf War in 1991.[7]

Vietnam presented the most vexing challenge for US national values. President Kennedy was the most articulate of the four US presidents who managed foreign policy during the Vietnam era. He saw the conflict 'not so much a military as a political and moral challenge' and believed that through nation-building the US could strengthen the South Vietnamese to resist communism. Nation-building was the predecessor of nation assistance, and emphasized mobilizing public support for military and political legitimacy through civic action and domestic reform.[8]

According to Kissinger, US support for the overthrow of President Diem in 1963 'cast its involvement in Vietnam in concrete' by committing the US to fill the political vacuum with legitimate successors:

> Ultimately, every revolutionary war is about governmental legitimacy; undermining it is the guerrillas' principal aim. Diem's overthrow handed that objective to Hanoi for free … [I]n the end, *legitimacy* involves an acceptance of authority without compulsion; its absence turns every contest into a contest of strength. [emphasis added][9]

Vietnam produced its share of lessons learned in legitimacy, which will be discussed in Chapter 6. Once US ground forces were committed to combat in 1965 the die was cast; nation-building became warfighting, and any residue of moral values was lost in the pervasive ambiguity of a war without a clearly defined enemy. The US could not provide security for the Vietnamese people against relentless attacks from the Viet Cong and later the North Vietnamese army; and collateral damage caused by US offensive operations exacerbated the resentment of the Vietnamese people toward the US and a corrupt South Vietnamese government. By 1968 the moral highground had been lost, as evidenced by the erosion of public support in America.

Kissinger succinctly summarized the US dilemma, which was to ignore the requirements and principles of military and political legitimacy:

> Too idealistic to base its policy on national interest, and too focused on the requirements of general war in its strategic doctrine, America was unable to master an unfamiliar strategic problem in which the political and military objectives were entwined. Imbued with the belief in the universal appeal of its values, America vastly under-estimated the obstacles to democratization in a society shaped by Confucianism, and among a people who were struggling for political identity in the midst of an assault by outside forces.[10]

In 1993 a similar dilemma would be revisited in Somalia, and the results would be similarly tragic, albeit on a smaller scale.

Kissinger brings US diplomacy up to date with a reaffirmation of the national values of democracy, human rights and the rule of law to US foreign policy, applied with a touch of *realpolitik*. President Carter championed human rights as an element of diplomacy, but so did Presidents Reagan and Bush. Human rights helped end the Cold War: the so-called 'Basket' of human rights established by the Helsinki Agreement of 1975 was ultimately instrumental in toppling communist regimes in Eastern Europe.[11]

The history of diplomacy has taught that the national values of democracy, human rights and the rule of law must be considered broad and flexible concepts with universal application, not limited to ethnocentric cultural standards drawn exclusively from US experience. Their unique relationship must also be understood. Democracy, human rights and the rule of law are inextricably bound together; none can stand alone and fulfill the requirements of legitimacy. Democracy can be tyrannical if not coupled with the protection of minority human rights through the rule of law. But too much emphasis on the rights of individuals or groups can defeat the legitimate (and essential) collective interests of the state. And the rule of law can be tyrannical if its purposes are subverted to causes other than preserving democracy and human rights.

The remainder of this chapter considers democracy, human rights and the rule of law as strategic concepts that frame issues of political and military legitimacy, and illustrates why these libertarian values must balance the authoritarian values prevalent in the military. Promoting these national values is implicit in operations other than war, and involves both military-to-military and civil-military operations. These values are especially important in emerging democracies where militaries have a tradition of human rights abuses. Achieving US political objectives in such environments requires military leadership that understands the importance of the requirements of military legitimacy to mission success – leadership that can function as an extension of both the military and diplomatic corps.

DEMOCRACY

The end of the Cold War was proclaimed as the victory of democracy over communism. But democracy has not always been seen as superior to authoritarianism. Writing in 1942, Quincy Wright blamed the isolation and appeasement policies of democratic governments for the two World Wars. Inherently peaceful and conciliatory, democracies were slow to deter the aggressions of more authoritarian regimes:

> Democracy has stimulated the will of the people to eliminate war, although it has not yet enlightened their intelligence as to the means.[12]

In the wake of the Cold War, much of Wright's reasoning is again relevant to the role of democracies in a violent world. The inability of democracies and their international organizations, such as NATO and the UN, to deal with contemporary violence and aggression, such as that in Bosnia, could again lead to wider wars. Ironically, it is a democratic principle – self-determination – that feeds this violence.

The demise of democracy?

The English statesman Edmund Burke once warned Americans that they would ultimately forge their own shackles. Troubling trends in the US indicate that Americans may be doing just that. But the malaise of democracy is not limited to the US.

Three criteria have been postulated for the demise of democracy around the world: historical dislocation, disaffection with all political leadership, and skepticism about social progress. There has been a tendency for emerging democracies, especially those with an authoritarian past, to disintegrate into tribal, ethnic, and religious conflict when petty despots appeal to intolerance, fear, and hate to mobilize support for violence. That violence, termed by Charles Maier as territorial populism, can evolve into ethnic cleansing. To counter such violence Maier recommends that governments promote democracy beyond borders through coalition-building.[13]

Individual vs group rights

Civil war can result when groups put territorial, tribal, ethnic, racial, or religious loyalties ahead of national loyalties. Today, as 130 years ago, a house divided against itself cannot stand. Urban riots continue to remind Americans that ethnic and racial polarization remain a threat to internal security.

A stable democracy must balance the rights of groups to self-determination with the protection of individual rights. Unless minority rights are protected by law, democracy can lapse into a tyranny of the majority. This has been evident in theocracies that have eliminated individual rights with unyielding religious law.[14] Algeria, Egypt, Saudi Arabia and Kuwait could become fundamentalist Islamic republics like Iran and Sudan if their majorities were allowed to choose their political future.[15]

Western democracies have prevented violent polarization through a careful balance between the rights of individuals and the collective rights of society as a whole. The US Constitution provides that balance: the Bill of Rights defines fundamental individual rights and the Fourteenth Amendment strengthens individual rights by guaranteeing to everyone 'equal protection of the laws'. This goal of the Constitution is represented by the scales of justice and proclaimed on the portico of the Supreme Court: Equal Justice Under Law.

Over the years individual rights have been expanded by civil rights laws that have prohibited discrimination based on race, ethnic origin or religious beliefs. But a subtle change has occurred that has significant implications for the future: civil rights laws have shifted their focus from protecting individual rights to protecting the rights of designated classes or groups, including ethnic and racial minorities. Rather than moderating ethnic and racial differences, such laws have exacerbated them.

Special interest groups have been successful in obtaining civil rights legislation that gives them special protection, and in some cases special preferences as compensation for past inequities (for example, Native Americans and Blacks). The shift in civil rights legislation from individual to group rights is one of the troubling trends mentioned in Chapter 2, and reflects political polarization along racial and ethnic lines. It has also clouded constitutional priorities for individual and collective rights. Ironically, individuals suffering discrimination who cannot identify with a protected class may have difficulty finding relief under current civil rights laws.

Self-determination: the ultimate group right

The ultimate right of any group within a society is self-determination. The right to dissolve an existing political union is a fundamental principle of democracy, but one often associated with violence. Since the American Revolution the concept of self-determination has represented freedom and individual rights to Americans, but for demagogues in former communist regimes it has provided a rationale to destroy old unions and legitimize ethnic cleansing.

There is an irony here: during the Cold War the West promoted the individual values of democracy over the collective values of communism. But despite the victory of democracy over communism that ideal has been subverted. In former communist countries such as Bosnia violent self-determination has come at the expense of human rights; and in others the dark side of democracy – liberty turned to license – has many supporting the return of authoritarianism to restore law and order.

There is considerable inconsistency, if not hypocrisy, in the US quickly recognizing the dissolution of the Soviet Union and Yugoslavia. It is true that the American Revolution was a violent act of self-determination, but the War Between the States reversed US policy on self-determination. The attempt of the southern states to secede from the Union was similar to the declaration of independence of the 13 original states; but President Lincoln condemned such self-determination, pronounced the Union to be sacrosanct, and committed US military force to preserve it at all costs. After more than 600,000 deaths the prohibition against self-determination was written in American blood.

The dissolution of the Soviet Union and Yugoslavia has also been associated with bloody conflict. A revival of nationalistic populism suppressed during the Tito years has resulted in primal conflict that has made a mockery of human rights.[16] If human rights are to mean anything the perpetrators of such war crimes must be brought to justice. Gidon Gottlieb has noted:

> The war on genocide should, at a minimum, mean that those who commit genocide – and are formally indicted for the crime – shall never rest, that they shall enjoy neither immunity nor protection. They should have the legal status of outlaws, subject to seizure, just as pirates were for centuries. Their properties and financial assets should be frozen everywhere.[17]

International enforcement of fundamental human rights could discourage ethnic and religious minorities from resorting to civil war to establish their independence. A generic bill of rights for emerging democracies would be an even better deterrent. Once individual rights and political redress are embedded in the fabric of government, disenchanted minority groups are not likely to resort to violent self-determination.[18]

Self-determination as a domestic threat

While current attention is focused on overseas violence, the US cannot overlook its own troubling trends which are summarized in Chapter 2. These trends could develop into serious threats if polarization continues and ethnic, racial or religious groups begin to coalesce and advocate self-determination. The fact that some already refer to themselves as black nationalists is cause for concern.

Self-determination should be discouraged in the US to promote peace and stability. If political sub-divisions such as cities, counties or states should ever become racial, ethnic or religious enclaves, the result could be as destructive of the national fabric as was secession. Such a scenario is conceivable in the US: the proliferation of racially defined single-member legislative districts and the polarization of partisan politics along racial lines could lead in that direction.

The best way to discourage racial and ethnic groups from seeking independent political enclaves is to abandon the concept of group rights and political districts for racial and ethnic minorities. A return to individual rights balanced against the collective rights of the state is the best insurance of equal justice under law, a guiding principle of US democracy that should not be confused with the right to be made equal by the law.

The two objectives of minimizing discrimination against racial and ethnic minorities and discouraging self-determination require rebuilding coalitions among increasingly adversarial special interest groups. History has taught that assimilation, not polarization, is the best route to justice for ethnic minorities in America.[19]

Stillborn democracy

There is another threat to democracy found in societies with a tradition of ruling élites. In Africa and Haiti opportunities for democracy have been stillborn because populations have been unable or unwilling to resist militant ruling élites. Post-colonial experiments with democracy in Africa have evolved from autocracy to anarchy. Haiti is unique in that it does not have tribal loyalties to complicate matters, but even the dramatic return of President Aristide under the protection of US forces is no guarantee of lasting democracy there.

Military intervention to establish democracy is impractical when a population is unwilling or unable to assume the responsibilities of self-government, which includes defending democracy against internal as well as external threats. The US military intervened in Haiti in 1915 and remained for 19 years with little effect. European powers had similar experiences in Africa. As the US recently learned in Somalia, unless a population assumes the responsibilities for self-rule military intervention is not likely to promote democracy but instead be perceived as neo-colonialism. Whether the 1994 intervention in Haiti is counted a success or failure will depend upon whether the Haitian people assume the responsibilities of defending democracy, human rights and the rule of law against those who will certainly challenge these values after the multinational forces leave.

HUMAN RIGHTS

Just War and human rights

When in conflict, the protection of human rights takes precedence over self-determination. Where primal violence results in genocide, military intervention may be necessary to restore the rule of law and protect fundamental human rights. What are these human rights that can justify military intervention?

The concept of human rights in wartime developed out of the Just War tradition.[20] That tradition has imposed few restrictions on how combatants attack each other; after all, that is what war is all about. But civilians are another matter. Since the Middle Ages, when rape, plunder and pillage ceased to be tolerated as legitimate spoils of war, Just War has placed increasing emphasis on protecting the lives and property of civilians from the ravages of war. Its most important standards have been codified in the law of war, which is discussed below.

Just War principles go beyond the law and beyond war. They represent moral and ethical constraints on the use of lethal force, even if they have been frequently disregarded. The principles of Just War complement the law of war, and provide moral guidelines for the use of military force where laws are inadequate. They represent the convergence of law, morality and values – the elements of military legitimacy.

In peacetime, Just War principles are especially important to strategic and operational restraint. When there is no threat to vital US interests, these strategic principles provide moral justification for military intervention; and when intervention is justified, Just War principles provide moral guidelines for the use of force at operational and tactical levels.

Wartime is different. It is a lethal contest that contemplates death and destruction. In order to protect non-combatants from the ravages of war the law of war distinguishes them from combatants, who are legitimate targets in war. Only force calculated to cause unnecessary suffering is prohibited against enemy combatants; that is, unless and until they become non-combatants through incapacity or surrender, when they are entitled to the same protections afforded other non-combatants.[21]

In operations other than war the distinction between combatants and non-combatants is blurred. The enemy, if one can be identified at all, can be an innocent civilian by day and a ruthless combatant by night.[22] The questionable status of armed youngsters in Somalia illustrated this ambiguity; they were dangerous to UN forces whether they were combatants or bandits. Such ambiguities make the law of war an inadequate standard of legitimacy in operations other than war.[23] Rules of engagement (ROE) tailored to specific mission requirements are necessary to provide the standards of restraint required for military legitimacy.

Sovereignty: national rights vs human rights

At the strategic level the moral and ethical standards for going to war originated with religious concepts (the principles of just cause and right intention), but they were secularized in the seventeenth century when Hugo Grotius published his treatise, *The Law of War and Peace* (1625). Grotius justified war based on self-defense, protection of property, state duties and punishment for aggression. Moral concepts of just cause and right intention became less important than the sovereign right of nations to protect their interests. With the acceptance of the doctrine of national sovereignty in the eighteenth century, non-interventionism, or war avoidance, was the moral standard; but as discussed earlier, the rule was honored in its breach, confirming that might made right.

The legal doctrine of sovereignty considers each nation equal and independent – free to resolve its own internal problems however it chooses, no matter how brutal or at what cost of human rights – without risk of intervention. This principle of non-intervention has been codified in Article 2 (4) of the United Nations Charter, which prohibits '. . . the threat or use of force against the territorial integrity or political independence of any state, or in any other manner inconsistent with the purpose of the United Nations'. Only two exceptions are provided for: enforcement actions ordered by the Security Council pursuant to Article 42 (used in *Desert Shield/Storm*) and individual and collective defense against aggression as provided in Article 52.[24]

The principle of non-intervention has produced creative theories based on Just War principles that have expanded the concept of self-defense. Strict interpretation of Article 52 requires an armed attack before armed force is justified in self-defense. Just War principles were used to stretch the concept of self-, or collective, defense in order to justify US interventions in Grenada in 1983 (*Urgent Fury*) and Panama in 1989 (*Just Cause*). These justifications included the protection of US citizens and the request of putative governments for military assistance. The US bombing raid on Tripoli in 1986 went even further, having been justified as pre-emptive self-defense. This same rationale was used by the Israelis in their attacks on PLO bases in Arab countries.[25]

Operation *Restore Hope* established a precedent for military intervention under the auspices of the UN without either an invitation or grounds for individual or collective self-defense. The only justification was human suffering caused by civil (tribal) violence and aggravated by famine and over-population. Somalia was not unique; Rwanda has since experienced even greater carnage, and Angola and Sudan are experiencing similar primal conflict. Given its unhappy experience in Somalia the US is not likely to intervene in African violence for purely humanitarian reasons; but *Uphold Democracy* in Haiti illustrated that the Somalia precedent could be applied closer to home.

Changing concepts of sovereignty

The erosion of the traditional concept of sovereignty as a bar to intervention has sparked interest in creating a UN peace-enforcement capability that could complement its traditional peacekeeping role. Secretary General Boutros Boutros-Ghali has proposed such an offensive capability while acknowledging that 'fundamental sovereignty and the integrity of the state remain central'. He went on to advocate a more relative concept of sovereignty:

> ... sovereignty was in fact never so absolute as it was in theory. A major intellectual requirement of our time is to rethink the question of sovereignty.[26]

He has also proposed a novel concept of universal sovereignty that would support offensive peace operations and humanitarian intervention under international law:

> Underlying the rights of the individual and the rights of peoples is a dimension of universal sovereignty that resides in all humanity and provides all peoples with legitimate involvement in issues affecting the world as a whole. It is a sense that increasingly finds expression in the gradual expansion of international law.[27]

Boutros-Ghali's concept would impose on the community of nations an affirmative obligation to intervene wherever there are gross violations of human rights or crimes against humanity. The prospect that the UN may become the world's policeman, with the coercive capability to enforce its decisions, has made many UN members nervous about the crumbling shield of sovereign immunity. Democracy has been recognized as a universal ideal for each nation, but not for the community of nations (the UN), where the fiction of sovereignty (one nation, one vote) would put inordinate power in the hands of the Third World.

Gidon Gottlieb has proposed another unique concept that would modify the concept of sovereignty in order to discourage ethnic violence. He has suggested the creation of extra-territorial national home regimes that would provide protective zones for ethnic minorities.[28] These international ethnic zones should be distinguished from internal ethnic political sub-divisions that threaten the stability of existing nation-states.

Gottlieb has made a distinction between nations (ethnic groups seeking national autonomy) and states (internationally-recognized entities) to support the concept of national home regimes. His theory is that if ethnic groups such as the Kurds, whose traditional homeland crosses several national boundaries, have legally-protected safe havens (home regimes), they will be less likely to resort to violence to create their own states.[29]

An underlying premise of Gottlieb's theory is that only democratic regimes are legitimate, but that self-determination is at the root of contemporary ethnic violence in much of the world.[30] He supports military intervention to alleviate the human suffering associated with violent self-determination, and advocates humanitarian assistance and the special military capabilities to provide it; he has even suggested that Special Forces units be used to apprehend those accused of war crimes. But he has also recognized the limitations of combat power, citing the Weinberger Doctrine and suggesting that military enforcement operations should rely on air power, not US ground forces.[31]

Humanitarian intervention as Just War

Eroding concepts of sovereignty have made the legal principle of non-intervention less absolute, restoring the relevance of the Just War principles to intervention. As the traditional theory of sovereignty gives way to moral justifications for intervention, current legal standards of self-defense and collective defense are likely to be broadened for humanitarian intervention.[32]

The three major Just War principles governing intervention are competent authority, just cause and right intention. Competent authority is a legal standard governed by the Constitution and statutory law such as the War Powers Act, and international law such as the United Nations Charter. But just cause and right intention are moral criteria that recall the theological underpinnings of Just War.

Just Cause applies the principle of discrimination to strategic decisions to intervene. Considerations include

> ... the substance of the cause, the comparative justice of the adversaries, the proportionality of the means and consequences of recourse to armed force to the good to be achieved, in the light of the probability of success and reasonable exhaustion of peaceful remedies.[33]

Right intention has theological origins, and includes promoting values such as democracy, human rights and the rule of law. Right intention is by nature more abstract than just cause, but would prohibit any military means beyond that necessary to achieve a just peace.[34] Right intention applies the principle of proportionality to strategic decisions regarding the kind, degree and duration of military force used in operations other than war.

The new interventionists

The revival of moral standards to justify military intervention has brought new support for humanitarian intervention. Theologians and church groups that have traditionally opposed military interventions are now echoing Secretary General Boutros-Ghali's call for new interpretations of sovereignty and military legitimacy:

> People are calling for reinterpretations of the concept of both national sovereignty and non-intervention, saying that the way we've understood them for three centuries is not adequate.[35]

The Reverend William Sloane Coffin, Jr., one of the most outspoken critics of US military operations since Vietnam, supported the US intervention in Somalia and has indicated a willingness to support peacekeeping in Bosnia:

> Moral isolation is simply not a defensible position for those opposed to war. There is great anguish and confusion. We are groping for some kind of legitimate police action on an international scale.[36]

Rev. Sloane expressed the sentiment of many who have had misgivings about the morality of military operations in the past, but who feel morally obligated to use the military instrument of national power to protect fundamental human rights. This was reflected in a rare joint statement of leaders of major Protestant, Catholic, Jewish and Muslim groups made before the deployment of US forces to Somalia. The resolution confirmed that the US '... is not policeman to the world, but the mass murder of innocents is unacceptable'. It went on to say that the US should '... act in concert with other nations when possible, alone when necessary'.[37]

Michael Walzer, a recognized authority on Just War, believes that moral standards incorporated in the principles of just cause and right intention are sufficient to justify intervention in Bosnia:

> I think of this in terms of the old international doctrine of humanitarian intervention. It was always held that in cases of massacre on the other side of the border, you have a right, and maybe an obligation, to go in and stop it if you can. I think that applies to starvation, whether politically induced or naturally caused, or to ethnic cleansing, mass deportations, and other acts that, in the old legal phrase, 'shock the moral conscience of mankind'.[38]

But Walzer warns of using moral pretensions for immoral purposes, the essence of wrong intention. He noted that a crucial condition of a right intention to support military intervention is that it not be '... a cover to create a satellite state, a puppet government, or be used for conquest'.[39]

The new dialogue on Just War indicates a major shift in public support for humanitarian intervention that is likely to effect future US military commitments. Ironically, many conservatives who have supported military interventions in the past are opposed to humanitarian interventions as an inappropriate use of the US military. The debacle in Mogadishu temporarily chilled support for US interventions; but the initial public support for *Restore Hope* and more recently for *Uphold Democracy* in Haiti portends a humanitarian role for the US military in peacetime – a capability seen by many as a moral imperative rather than a security measure.[40]

The revival of just cause and right intention as moral justifications for intervention reflects a circular evolution of Just War. Almost 400 years after Grotius began the secular trend with the doctrine of sovereignty, the law has proved to be an inadequate substitute for morality. In the new security environment both legal and moral standards are essential elements of military legitimacy.

The importance of just cause and right intention to legitimize any uninvited US military intervention was reflected in the UN and congressional resolutions authorizing *Desert Storm*. The congressional resolution authorized offensive action based on the requirements of these Just War principles.[41] The circumstances surrounding the UN resolution which authorized *Restore Hope* were different in that there was no external agression; the legal basis for the intervention was disaster relief.[42] In Haiti, the acquiescence of General Cedras, even if under duress, and subsequent congressional resolutions legitimized the US military intervention there.

THE RULE OF LAW

Human rights in wartime: the law of war

As mentioned earlier in this chapter, the law of war addresses human rights with rules that have evolved from the Just War tradition. Some of these rules have become part of customary (unwritten) international law and the law of war through usage and custom over the years. They include the principles of military necessity, discrimination and proportionality that are part of the customary law of war, as is the principle of humanitarian treatment.[43]

More specific standards are provided by treaties or international conventions. The four Geneva Conventions of 1949 and the UN Charter are among the most important sources of the law of war on human rights. They define war crimes and provide specific standards for treating civilians and other non-combatants; they elaborate but do not change the standards first codified in the Lieber Code of 1863.[44]

The focus of the Geneva Conventions is on protecting civilians and other non-combatants from the violence of warfare. Combatants forfeit their right to protection against lethal force when they put on a uniform and become a lawful target for enemy combatants. But even combatants are protected against weapons calculated to cause unnecessary suffering, such as dumdum bullets, poisons and bacteriological weapons. The law of war also prohibits treachery among combatants, which includes the improper use of a protected symbol, such as a red cross or a white flag of truce, to gain advantage over the enemy.[45]

The law of war provides protection for non-combatants on the theory that war should not harm those who do not make war. Under the Geneva Conventions non-combatants such as civilians, prisoners-of-war, and the sick and wounded, are considered protected persons and entitled to protection from inhumane treatment. Violations are considered war crimes. Combatants are not protected unless wounded or captured, when they become non-combatants.

War crimes defined in the Geneva Conventions fall into two categories: grave breaches, which are subject to severe punishment, and other war crimes considered less serious. Grave breaches include 'willful killing, torture, or inhuman treatment, including biological experiments, willfully causing great suffering or serious injury to body or health', as well as serious property crimes, including the 'extensive destruction and appropriation of property, not justified by military necessity and carried out unlawfully and wantonly'.[46] Parties to the Conventions, including the US, are required to enact laws to punish those committing grave breaches, while they must only *suppress* other unlawful acts.

The *Geneva Convention Relative to the Protection of Civilians in Time of War* illustrates the special protected status of civilians:

> Protected persons are entitled, in all circumstances, to respect for their persons, their honor, their family rights, their religious convictions and practises, and their manners and customs. They shall at all times be humanely treated, and shall be protected especially against all acts of violence or threats thereof and against insults and public curiosity. Women shall be especially protected against any attack on their honor, in particular against rape, enforced prostitution, or any form of indecent assault.[47]

The Geneva Conventions apply to armed conflicts between two or more nations; but even in civil wars, like the one in Bosnia, which are not covered by the full Geneva Conventions, non-combatants are entitled to protection. In addition to the protection afforded by customary law, non-combatants

> ... shall in all circumstances be treated humanely [and] ... the following acts are and shall remain prohibited at any time and in any place: ... violence to life and person ... [and] outrages against personal dignity, in particular, humiliating and degrading treatment.[48]

Contemporary war crimes

During *Desert Shield/Storm* it was evident that grave breaches of the Geneva Conventions were committed by Iraqi forces in Kuwait. There has been even more evidence of war crimes in Bosnia. The failure of the world community to bring these war criminals to justice illustrates the weakness of international law: there is no effective enforcement mechanism.

The UN Security Council has convened a war-crimes tribunal in The Netherlands; but there has been no other international tribunal with compulsory jurisdiction to prosecute war crimes since Nuremberg. The US has not helped to fill this void; it even used the weakness of the existing International Court of Justice, which lacks mandatory jurisdiction, to its advantage when it refused to appear before that Court after charges were brought against the US by the Nicaraguan government (then controlled by the Sandinistas) for supporting the Contras.[49]

Individual states have jurisdiction to try war crimes, however. In March 1993 a Bosnian military court condemned to death two Serb soldiers convicted of war crimes. They were accused of killing at least 40 people, many of them young women who were first raped as part of the Serb campaign of ethnic cleansing. Ironically the verdict was condemned by the commander of UN forces in Bosnia; French General Philippe Morillon criticized the Bosnians for trying the Serbs by their own military court rather than waiting to prosecute them before the UN war crimes tribunal.[50]

Rape and murder have apparently been commonplace in the Bosnian civil war. A report by Amnesty International, *Bosnia-Herzogovenia: Rape and Sexual Abuse by Armed Forces*, states 'The available evidence indicates that in some cases the rape of women has been carried out in an organized or systematic way, with the deliberate detention of women for the purpose of rape and sexual abuse.' While most of the reported rapes have allegedly been committed by Bosnian Serbs against Moslems, incidents have been reported on all three sides of the conflict: Serbs, Croats, and Moslems. All have blood on their hands.

While history provides examples of rape in almost every war, the Bosnian civil war is unique in that Serbian 'ethnic cleansing' policy apparently includes the rape of Moslem women. And to make it worse, the policy seems to be achieving its objectives: to intimidate and humiliate Bosnian Muslims into leaving areas claimed by Serbs. Rape and enforced pregnancies have contaminated Muslim concepts of identity and nationality, which are based on ethnic purity. The intentional use of rape as an instrument of war has caused the Balkans to become 'a sort of Bermuda Triangle into which human decencies vanish without a trace'.[51]

Since the creation of the UN Yugoslav War Crimes Tribunal, twenty-one Serbs have been indicted for atrocities committed at a military prison camp. But only one, who happens to be in German custody, is likely to face trial. The rest, along with Serbian leaders President Slobodan Milosević and Radovan Karadzić who were named by former Secretary of State Eagleburger as possible war criminals in 1992, remain at large.

If the new UN War Crimes Tribunal cannot enforce the law of war against those who violate it with such impunity, the law will cease to be a meaningful standard of legitimacy. The indictments are the first on genocide since the trials of Nazi war criminals at Nuremburg after the Second World War. The difference is that the Allies had custody of those indicted then, while the UN seems powerless to do the same today. The practical uselessness of the indictments underscores the ineffectiveness, so far, of the UN to enforce the most fundamental of human rights.[52]

The universality of jurisdiction over war crimes allows any nation to prosecute war criminals. This supports the legality of the Bosnian court as well as the UN War Crimes Tribunal. While the enforcement of international law by the courts of nations at war (or sympathetic to one party or the other) may complicate peace negotiations, the law should not become a casualty of war in order to expedite peace. For those in the US military, jurisdiction is not an issue; for them, war crimes are offenses under the Uniform Code of Military Justice and punishable by court-martial.[53]

Human rights in peacetime

As discussed earlier, the egregious violation of human rights is justification for military intervention. What are these human rights that pre-empt sovereignty? While the Geneva Conventions define fundamental human rights in international armed conflict and civil war, there is no peacetime equivalent. The 1977 Geneva Protocols which would have extended the definition of war to civil conflicts were never ratified by the US. Nevertheless, US policy has extended the highest standards of human rights in the Geneva Conventions to all conflicts in which its forces have been involved.

Beyond the Geneva Conventions the definition of human rights and the collective enforcement mechanisms in peacetime are found in treaties to which the US is a party. The most important of those collective arrangements are the UN and Organization of American States (OAS).

The UN Charter provides in Article 1(3) that one of its principal purposes is '... promoting and encouraging respect for human rights and for fundamental freedoms'. Article 55 provides '... the UN shall promote ... universal respect for, and observance of human rights and fundamental freedoms for all without distinction as to race, sex, language, or religion'. Article 56 pledges all members '... to take joint and separate action ... for the achievement of the purposes set forth in Article 55'. But the Charter fails to define peacetime standards for human rights or provide effective mechanisms to enforce them. The lack of standards and enforcement mechanisms to hold offenders accountable is a glaring deficiency, not only with regard to atrocities in Bosnia, but also in Latin America and Haiti.

The OAS Charter suffers the same weakness as the UN Charter. In Article 3(j), OAS members '... proclaim the fundamental rights of the individual without distinction as to race, nationality, creed, or sex'. And in Article 16 each member pledges to 'respect the rights of the individual and the principle of universal morality'. These vague standards and the lack of effective enforcement mechanisms have made human rights almost meaningless in Latin America and Haiti. This is evident in the 1993 UN Truth Commission Report which cited egregious human rights violations in El Salvador and continuing human rights violations in Haiti. With no enforcement mechanism, little action is likely to be taken to remedy these wrongs.

The Universal Declaration of Human Rights is not a treaty, but it was adopted by the UN General Assembly in 1948 as an authoritative interpretation of the Charter's general requirement to promote respect for human rights. The American Declaration of the Rights and Duties of Man was adopted by the OAS in a similar fashion. Both Declarations set forth specific rights which reflect international principles and US national values: respect for the individual, democratic institutions, and rule of law. The State Department views the Universal Declaration as 'the most important and widely-accepted standard-setting document in the world', and, along with the American Declaration, uses it to measure the human rights performance of states for its annual report to Congress.[54]

While not the law of the land, the standards of the Universal and American Declarations are US policy and are incorporated into rules of engagement (ROE) and general orders which govern the conduct of US forces overseas. The Declarations define human rights in the context of democracy and the rule of law, and provide meaningful if not enforceable standards for military legitimacy. While these standards affect the legitimacy of all military operations and activities, they are primary requirements of military legitimacy in those security and humanitarian assistance activities discussed in Chapter 2.

Operational law and human rights

The law of human rights is a part of operational law, or OPLAW. OPLAW is the rule of law applicable to the military;[55] its foundation is the Constitution, and its components are US domestic law, regulations and directives, international law and host country laws. The most important standards of OPLAW relate to restraints on the use of force: the law of war provides the standards of restraint in war, and the more restricted standards required for operations other than war are provided by ROE. OPLAW also provides additional restrictions on operations other than war that are discussed in Chapter 2 as they relate to specific activities.

The standards of OPLAW that relate to the use of force and ROE are analogous to the moral standards of Just War and govern the operational and tactical issues of legitimacy. These standards of legitimacy are tailored to the unique military and political objectives of operations other than war through ROE, which are discussed in Chapters 3 and 6 as elements of restraint.[56]

Both OPLAW and peacetime ROE emphasize the importance of human rights, especially in operations other than war. If there has been a crucible for testing OPLAW and ROE in the last decade, it has been Latin America. There human rights violations were allegedly committed by officers trained at the US Army School of the Americas at Fort Benning, Georgia, raising questions about the school's commitment to human rights. Emphasizing the priority of human rights training at the school, General Barry R. McCaffrey, Commander of US Southern Command (USSOUTHCOM), has cited the OAS charter and the American Declaration of Human Rights as an expression of OPLAW standards which must be respected by all military officers in Latin America. General McCaffrey confirmed the linkage between human rights, democracy, and the rule of law, citing John Shattuck, Assistant Secretary of State for Human Rights and Humanitarian Affairs:

> Human rights, democracy and the rule of law are not the same. But they are complementary and mutually reinforcing ... Democracy – the rule of, by, and for the people – is only possible in a political and social order that fully respects the rights of each and every man, woman, and child in society. . . . Governments that do not respect the rule of law are by definition lawless.[57]

As an OPLAW standard of culpability, General McCaffrey cited the *Medina* standard, used in the prosecution of those involved in the My Lai massacre in Vietnam:

> If a captain, colonel, or general knows of a human rights violation or war crime, and takes no action, then he or she will be held criminally liable. That's what we teach everyone here at this institution, at the School of the Americas.[58]

To illustrate how respect for human rights – a variation of the golden rule – is incorporated into military leadership, General McCaffrey cited an example discussed in Chapter 1: He contrasted the calculated brutality of General William T. Sherman toward Southern civilians with the respect shown by General Robert E. Lee toward Northern civilians. It was a classic example of how winning the war, when it involves the abuse of civilians and their property, can come at the expense of losing the peace:

> Winning a war is a reasonably easy proposition. It involves energy, courage, violence, and organization. Winning the peace is a far more difficult thing to do. General Sherman's actions, his barbarity and cruelty, created a hundred years of bitterness in the American South; some aspects of which endure today. General Lee, on the other hand, consistently espoused values [treating civilians and their property with respect] which were not and are not a military weakness.[59]

General McCaffrey summarized the concepts of democracy, human rights, and the rule of law into seven principles of legitimacy and leadership that are a preamble to the next chapter:

- zero tolerance for (human rights) abuse
- human rights training
- understanding ROE
- treating soldiers (and civilians) with respect
- lead by example
- control your troops
- honorable conduct pays off[60]

SUMMARY

The history of US diplomacy reflects the evolution of the national values of democracy, human rights, and the rule of law as core principles of US foreign policy. These constitutional values have a moral context, but in the strategic application of military force they must be tempered by the practicality of *realpolitik*.

Issues of democracy are complicated by the eroding concept of sovereignty, runaway self-determination, and the egregious violation of human rights. Innovative concepts of universal sovereignty and national home regimes offer alternatives to traditional sovereignty, but there is no substitute for the rule of law to protect the rights of minorities from the tyranny of the majority.

The Just War Tradition has provided moral principles relating to the use of force that complement national values. The Just War prerequisites for going to war are competent authority, just cause, and right intention; those requirements for warfighting are discrimination and proportionality. At the core of Just War is a moral principle inherited from the code of chivalry which is at the foundation of military legitimacy and civil-military relations: war should not harm those who do not make war.

The Geneva Conventions provide the primary legal standards for human rights in wartime, but peacetime standards are not so well defined. International standards for human rights in peacetime do not have the effect of law and there are inadequate enforcement mechanisms at the international level. Nevertheless, human rights are even more important to legitimacy in operations other than war than in warfighting.

The legitimacy of operations other than war requires that US military personnel understand the spirit as well as the letter of the law. That spirit is found in the Judeo-Christian heritage of our nation, best expressed in the golden rule. The professional values of duty, loyalty, integrity, and selfless service require an altruistic commitment to promote democracy, human rights, and the rule of law. The relevance of morality to military leadership will be further explored in the next chapter.

NOTES

1. Henry Kissinger, *Diplomacy* (New York: Simon and Schuster, 1994). Quincy Wright has cited the just war tradition and what he saw as a general trend toward humanism and tolerance (with total war a conspicuous exception) to illustrate how values have been an important ingredient in legitimizing warfare. Quincy Wright, *A Study of War* (Chicago: University of Chicago Press, 1942, 1971), pp. 157–158; 307–309.
2. Ibid. at p. 65.
3. Ibid. at p. 64.
4. Ibid. at p. 32.
5. Ibid. at p. 33.
6. Ibid. at p. 437.
7. Ibid. at p. 477.
8. Ibid. at pp. 648, 649.
9. Ibid. at p. 655.
10. Ibid. at pp. 698, 699.
11. Ibid. at pp. 759, 772.
12. Quincy Wright, *A Study of War*, n. 1 supra, pp. 4, 5, 839–848.
13. Charles S. Maier, 'Democracy and Its Discontents', *Foreign Affairs* (July/August 1994), pp. 48, 54, 61–63. For a related concept developed by Gidon Gottlieb, see infra n. 28.
14. An outline of Islam is provided in the appendix (pp. 303–306) to the *Operational Law Handbook* (JA 422, 1993) prepared by the Center for Law and Military Operations and the International Law Division at The Judge Advocate General's School, US Army, Charlottesville, VA (hereinafter *OPLAW Handbook*).
15. Judith Miller discusses the threat of Islamic fundamentalism to individual

liberty in 'The Challenge of Radical Islam', *Foreign Affairs* (Spring 1993), p. 43.

16. John Lukacs describes the proliferation of self-determination in Eastern Europe as 'populist nationalism', and considers it a danger to the human rights of ethnic and religious minorities: 'This is, of course, yet another manifestation of the potential tyranny of the majority – which as Tocqueville observed, is the great danger of democratic societies.' 'The End of the Twentieth Century', *Harpers* (January 1993), pp. 39, 54. See also articles by Huntington and Kaplan at n. 9 to Chapter 3.

17. Gidon Gottlieb, *Nation Against State* (New York: Council on Foreign Relations Press, 1993), at p. 124.

18. Robert Cullen has observed, 'The ultimate collective right ... is the right to create an independent state.' Cullen urges US foreign policy be 'focused firmly on individual rather than collective rights'. 'The Human Rights Quandary', *Foreign Affairs* (Winter 1992–93), pp. 79, 83–84. Quincy Wright cited Alexander Hamilton in *The Federalist # 31* on the tendency of state governments 'to encroach upon the rights of the Union' rather than the reverse. To equalize the tendencies of states to dissolve unions Wright argued that 'world authority must guarantee basic liberties within states'. Wright, supra n. 1, footnote on pp. 352–353; also pp. 909–911. Henry Grunwald has reaffirmed Lincoln's commitment to individual rights and the preservation of the union as the best safeguard for fair and stable government – Henry Grunwald, 'Memorandum to Woodrow Wilson', *Time*, 14 November 1994, p. 104.

19. In *Ethnic America*, Thomas Sowell has traced the history of major ethnic groups in America and concluded that discrimination diminished as ethnic groups assimilated into the larger society and that as long as racial or ethnic minorities maintained a separate (and adversarial) existence they were likely to be victims of discrimination. Thomas Sowell, *Ethnic America* (New York: Basic Books Inc., 1981).

20. Michael Walzer postulates life and liberty to be absolute values protected even in war at p. xvi and pp. 135 *et seq*. The protections of the Bill of Rights are discussed in Chapter 3. Michael Walzer, *Just and Unjust Wars* (New York: Basic Books, 1977), pp. 21 *et seq*. (hereinafter Walzer). Generally, for law of war standards, see FM 27–10, *The Law of Land Warfare* (July 1956, including Change 1 dated July 1976) (hereinafter FM 27–10); note the duplicity in para. 25 (p. 16) regarding civilians of an enemy nation: US law considers them to be the enemy while international law requires that they should not be the object of attack. On the tendency to ignore any distinction between combatants and civilians during total war (discussed in Chapter 1), see Quincy Wright, *A Study of War*, supra n. 1, pp. 305–310, 330–32, 810–12. The *DOD Law of War Program*, DOD Directive 5100.77 (July 1979) provides for the implementation of these wartime standards of military legitimacy, including the mandatory reporting of war crimes.

21. For a discussion of Walzer's just war principles applied to *Desert Storm*, see Yuval Joseph Zacks, 'Operation Desert Storm, A Just War?', *Military Review* (January 1992), p. 20. Zacks' conclusion that *Desert Storm* met the criteria for just war is questioned by Ranier H. Spencer in 'A Just War Primer', *Military Review* (February 1993), p. 20. Ranier questions whether the targeting of infrastructure that served both Iraqi civilian and military needs met the criteria of Walzer's revised 'double effect' standard. Elliot

Cohen would argue that it did. See Cohen, *The Mystique of US Air Power*, n. 47 to Chapter 1.

22. In 1977 two protocols to the Geneva Conventions were adopted to address the inapplicability of the law of war to contemporary conflict. Protocol I extended the definition of war (international conflict) to include 'armed conflicts in which peoples are fighting against colonial domination and alien occupation and against racist regimes in the exercise of their right to self-determination', and Protocol II provided the protections of the law of war to conflicts 'not of an international character'. See Geoffrey Demarest, 'Updating the Geneva Conventions: The 1977 Protocols', *The Army Lawyer* (November 1983), p. 18. See also n. 14, supra. John Jandora has described an armed populace as a threat in 'Threat Parameters for Operations Other than War', *Parameters* (Spring 1995), pp. 55, 59–61.

23. A military lawyer analyzing the application of the law of war to the US intervention in Panama concluded that the law of war has become '. . . too complex for practical application to the kinds of armed conflict that prevail today'. The author attributes the problem partly to '. . . biased perceptions of legitimacy of one side or the other . . .' See John Embry Parkinson, Jr., 'United States Compliance with Humanitarian Law Respecting Civilians During Operation Just Cause', *Military Law Review* (Summer 1991), pp. 31, 137, 138. Another criticism of the law of war during peacetime is that the rules '. . . are triggered only in situations of declared war or armed conflict situations that peacetime crises response is designed to avoid'. See Richard J. Erickson, *Legitimate Use of Military Force Against State-Sponsored International Terrorism* (Maxwell AFB, Alabama: Air University Press, 1989), p. 75.

24. Quincy Wright saw no inconsistency in simultaneously promoting human rights and state sovereignty, but then noted 'international law . . . has tended toward a recognition of absolute territorial sovereignty and abandonment of international standards for the protection of human rights . . .' Wright saw world public opinion as the ultimate sanction for international law that protects human rights, but acknowledged the conundrum that public opinion cannot develop in the absense of human rights. Wright, *A Study of War*, supra n.1, pp. 909, 911. As to *Desert Storm*, see Zacks and Spencer at n. 21, supra. For the legal bases for recent military operations, see the *OPLAW Handbook*, n. 14 supra at Tab D (pp. 54 *et seq.*).

25. For a discussion of 'preemptive self defense' by a former legal counsel to the State Department, see Abraham D. Sofaer, 'Terrorism, the Law, and the National Defense', *Special Warfare* (Fall 1989), p. 12. The article is also found in the Fall 1989 issue of *The Military Law Review*.

26. Boutros Boutros-Ghali, 'Empowering the United Nations', *Foreign Affairs* (Winter 1992–93), pp. 89, 98–99. Quincy Wright predicted changes in the concept of sovereignty, perhaps even its disappearance, in his 1942 book, *A Study of War*, n.1 supra, p. 921.

27. Ibid. at p. 99. A word of caution was given to Boutros-Ghali and other 'new interventionists' by Stephen John Stedman in 'The New Interventionists', *Foreign Affairs* (Winter 1992–93), Vol. 72, No. 1, p. 1. Stedman acknowledges the erosion of sovereignty but warns against a universal standard for humanitarian intervention in civil war, arguing that the UN is structurally incapable of performing that role. An overview of the changing nature of sovereignty is provided in the *Operational Law Handbook*, n. 14 supra at pp. W-253 *et seq*. On the interplay between sovereignty, international law

and human rights, see Quincy Wright, *A Study of War*, n. 1 supra, pp. 833–837, 907–922.

28. Gidon Gottlieb, *Nation Against State*, n. 17 supra. For a related concept, see n. 13 supra.
29. Ibid. at pp. 77–82.
30. Ibid. at pp. 20–24; see Grunwald at n. 18 supra. Martin van Creveld takes violent self-determination an additional step, viewing it as a trend toward anarchy. See Creveld, *A Transformation of War* (New York: The Free Press, 1991), Chapter 5.
31. Ibid. at pp. 114–116 (humanitarian assistance); pp. 117–121 (use of force).
32. See O'Brian, 'Special Operations in the 1980s: American Moral, Legal, Political, and Cultural Constraints', *Special Operations in US Strategy* (National Strategy Information Center Inc., New York, 1984), pp. 53, 59.
33. See O'Brian, unpublished paper entitled *Just War Doctrine's Complementary Role in the International Law of War* (1991), p. 16. Generally, on just war see Quincy Wright, *A Study of War*, n. 1 supra, pp. 155–162, 330–334, 385–387.
34. Ibid. at p. 22.
35. Peter Steinfels, 'Crises Altering Pacifists' Views', article in the *New York Times*, reprinted in *The State* (Columbia, SC), 21 December 1992, 1-A, 15-A.
36. Idem.
37. Idem.
38. Peter Steinfels, n. 35 supra, at 15-A.
39. Idem.
40. In an article entitled 'Wanted: A Golden Rule of Intervention' by Stephen Engelberg of *The New York Times*, Engelberg notes the absence of a national policy for humanitarian intervention and categorizes those on both sides of the issue: 'Gone are the traditional hawks and doves. In their place stand the new internationalists, who argue that the West has an overriding stake in encouraging order and should do so as a coalition wherever the conflict, and on the other side, the thinkers who oppose almost any intervention unless it can be justified by the familiar definitions of national interest. Engelberg quoted James Schlesinger as a pragmatist who saw the potential for humanitarian intervention but warned moralists to stay in touch with public opinion: 'Those who would act on behalf of morality had better think through the entire moral fabric including imposing those views on an indifferent body politic in America.' Reprinted in *The State* (Columbia, SC), 1 May 1994, p. A6.
41. The Joint Resolution passed by Congress (Public Law 102–01) on 14 January 1991, provided as follows:

> Whereas the Government of Iraq without provocation invaded and occupied the territory of Kuwait on August 2, 1990;
> Whereas both the House of Representatives ... and the Senate ... have condemned Iraq's invasion of Kuwait and declared their support for international action to reverse Iraq's aggression;
> Whereas Iraq's conventional, chemical, biological, and nuclear weapons and ballistic missile programs and its demonstrated willingness to use weapons of mass destruction pose a grave threat to world peace;
> Whereas the international community has demanded that Iraq withdraw uncon-

ditionally and immediately from Kuwait and that Kuwait's independence and legitimate government be restored;

Whereas the United Nations Security Council repeatedly affirmed the inherent right of individual or collective self-defense in response to the armed attack by Iraq against Kuwait in accordance with Article 51 of the United Nations Charter;

Whereas, in the absence of full compliance by Iraq with its resolutions, the United Nations Security Council in Resolution 678 has authorized member states of the United Nations to use all necessary means, after January 15, 1991, to uphold and implement all relevant Security Council resolutions and to restore international peace and security in the area; and

Whereas Iraq has persisted in its illegal occupation of, and brutal aggression against Kuwait: Now, therefore, be it

Resolved by the Senate and House of Representatives of the United States of America in Congress assembled . . .

Sec. 2(a) The President is authorized, subject to subsection (b), to use United States Armed Forces pursuant to United Nations Security Council Resolution 678 (1990) in order to achieve implementation of Security Council Resolutions 660, 661, 662, 664, 665, 666, 667, 669, 670, 674, and 677.

Sec. 2(b) Before exercising the authority granted in subsection (a), the President shall make available to the Speaker of the House of Representatives and the President *pro tempore* of the Senate his determination that –

(1) the United States has used all appropriate diplomatic and other peaceful means to obtain compliance by Iraq with the United Nations Security Council resolutions cited in subsection (a); and

(2) that those efforts have not been and would not be successful in obtaining such compliance.

Sec. 2(c)(1) Consistent with section 8(a)(1) of the War Powers Resolution, the Congress declares that this section is intended to constitute specific statutory authorization within the meaning of section 5(b) of the War Powers Resolution.

Sec. 2(c)(2) Nothing in this resolution supersedes any requirement of the War Powers Resolution.

Sec. 3 At least once every 60 days, the President shall submit to the Congress a summary on the status of efforts to obtain compliance by Iraq with the resolutions adopted by the United Nations Security Council in response to Iraq's aggression.

42. See *OPLAW Handbook*, n. 14 supra at V-241, 242.
43. Ibid. at p. Q-181. Quincy Wright has traced the evolution of the law of war, noting how humanitarian priorities were preempted by *realpolitik* and total war. Wright, *A Study of War*, n. 1 supra, pp. 305–310, 810–812, 909–911, and n. 33 supra. Martin van Creveld has cited the necessity for the law of war (the war convention) as primarily to benefit combatants by keeping civilians out of their way. Creveld, *A Transformation of War*, n. 30 supra, pp. 87–94.
44. Those provisions of the Lieber Code relating to the treatment of civilians are provided in n. 18 to chapter 1. For war crimes, see *OPLAW Handbook*, n. 14 supra, p. Q-183.
45. Ibid. at p. Q-182.
46. See FM 27–10 n. 20 supra, p. 179.
47. Ibid. at p. 106. But see the duplicity in FM 27–10 in the treatment of enemy civilians at n. 20 supra.
48. Ibid. at p. 9. In 1977, protocols to the Geneva Conventions were developed to broaden the meaning of war to include contemporary civil wars. Protocol I

extended the definition of war to include 'armed conflicts in which peoples are fighting against colonial domination and alien occupation and against racist regimes in the exercise of their right to self-determinism', and Protocol II provided the protections of the law of war to conflicts 'not of an international character'. Geoffrey Demerest, 'Updating the Geneva Conventions: The 1977 Protocols', *The Army Lawyer* (November 1983), p. 18. For the additional protections provided by Common Article III to Protocol II see, 'New Protection for Victims of International Armed Conflict', *Military Law Review* (Spring 1988), pp. 59–82.

49. For a discussion of the role of the International Court of Justice in enforcing international law, see Abraham D. Sofaer, *Terrorism, the Law, and the National Defense*, cited in n. 25 supra.

50. Article by David B. Attaway, *The Washington Post*, reprinted in *The State* (Columbia, SC), 31 March 1993, p. 1-A.

51. Article by Lance Morrow, *Time*, 22 February 1993, pp. 48–49.

52. Editorial, 'Tribunal underlies ironies', *The State*, 16 February 1995, p. A18.

53. See FM 27–10, at p. 182.

54. Human rights standards for US government personnel were established by State Department policy set forth in a November 1990 message relating to the Andean Initiative: 'Human rights, as defined by the UN and OAS Declarations and treaties and in US law include fundamental protections for the individual such as freedom from extrajudicial killing, torture, disappearance, and arbitrary arrest. Human rights as defined in these instruments also include civil rights essential to a democratic society such as the the the right to free expression, the right to assembly, the right to free and fair elections, the right to a fair trial, etc., as well as the right to an independent judiciary, and a government and military subject to the rule of law. These rights are the foundation of US human rights policy, which strives to protect the integrity of the individual and promote the democratic process, thus fostering peace and stability.'

55. Operational law has been defined as 'The body of domestic, foreign, and international law that impacts specifically upon US Forces in a combat and [peacetime] engagement operations.' *OPLAW Handbook*, n. 14 supra, at p. A-17.

56. See n. 33 to Chapter 3, supra.

57. Comments of John Shattuck (29 August 1993) cited by General Barry R. McCaffrey during his keynote address, 'The National Armed Forces as Supporters of Human Rights', at the US Army School of the Americas, Fort Benning, GA, 10 August 1994, p. 4.

58. Ibid. at p. 7.

59. Ibid. at pp. 6–7; the application of the golden rule is discussed in Chapter 1 (see n. 55 to Chapter 1 supra).

60. Ibid. at pp. 9–12; treating soldiers with respect incorporates the golden rule (see n. 55 to Chapter 1, supra).

5

Military Legitimacy and Leadership in Operations other than War

Better a patient man than a warrior,
A man that controls his temper than one who takes a city.

Proverbs 16:32

Whoever wants to be great among you must be your servant.

Matthew 20:26

Earlier chapters have provided an overview of the concept of military legitimacy and how it relates to operations other than war. This chapter has its focus on the relevance of military legitimacy to leadership in operations other than war. The civil-military focus of these operations requires a unique kind of military leader – one who can combine the proficiency of a combat leader with the finesse of a diplomat: the diplomat warrior.[1]

PARADOX OF THE MILITARY IN A DEMOCRACY

Some may consider the term diplomat warrior an oxymoron; it is true that diplomacy, traditionally a civilian endeavor, and military activities are not always compatible. But reconciling diplomacy with military operations other than war should be no more difficult than reconciling the paradox of a military organization within a democratic society, another prerequisite for military leadership.

Operations other than war require that the military leader be equally at home in a civilian or military environment. But there is a natural tension in civil-military relations that can affect military legitimacy. This tension is due to the tendency for military (collective) values to conflict with civilian (individual) values. Colonel Dennis R. Hunt, Professor of Law at the US Military Academy, introduces cadets beginning the study of Constitutional law to the potential conflict:

> To succeed as an officer you must comprehend the paradox of a military organization within a democratic society. The military is necessarily non-democratic and authoritarian, but it defends democratic principles and is manned with citizen soldiers drawn from a society which enjoys great personal liberties. You will be challenged to ensure that soldiers' Constitutional rights are neither unjustly nor unnecessarily abridged in the cause of accomplishing your mission and administering military law.[2]

The military environment emphasizes collective values such as good order and discipline which are required in an authoritarian organization; but these military values necessarily conflict with individual rights such as the freedom of expression which are protected by the Constitution. The potential conflict between these military and civilian values threatens civil-military relations and military legitimacy.

There is evidence that misplaced concepts of duty and loyalty to authoritarian values have caused some officers to lose touch with civilian values grounded in the US Constitution.[3] This was illustrated in the Iran-Contra affair, when Lieutenant Colonel Oliver North put his loyalty to a mission of doubtful legality ahead of his duty to support the Constitution. In failing to co-operate with the Congress, he not only compromised his integrity, but also his mission. His actions doomed congressional support for the cause he had so zealously pursued.

Colonel North's conduct has been described as a leadership failure that '... occurs when pragmatic but narrowly focused subordinates, in their zeal to get a job done or to please their boss, act illegally or unethically ...'.[4] A noted military ethicist, Colonel Anthony E. Hartle, questioned Colonel North's loyalty to the Constitution:

> Some critics have claimed of North that in his zeal to promote democracy abroad, he subverted it at home, specifically in subverting some of the fundamental tenets of the professional military ethic. North may have become so concerned about protecting foreign agents and contacts that he lost sight of his loyalty to American institutions and the Constitution.

Colonel Hartle noted that when Colonel North, or any officer for that matter, puts loyalty to mission ahead of loyalty to the Constitution, it is a real threat to democracy:

> When the inefficiency and lack of responsiveness of democratic procedures become too great a luxury or danger, and persons other than the people's elected representatives conclude that, because they understand the real priorities, democratic procedures must be set aside, then the republic is perhaps most endangered.[5]

The Constitution is the ultimate standard of legitimacy for military officers. They must not only understand the Constitution, but upon commissioning take an oath to 'support and defend the Constitution against all enemies, foreign and domestic; and bear true faith and allegiance' to the same. Where there are conflicting values of duty and loyalty, the oath leaves no doubt that the ultimate duty and loyalty of the military officer must be to the Constitution as the foundation of the rule of law and the bedrock of military legitimacy.[6]

The dichotomy between military and civilian values is most evident in emerging democracies where there has been no traditional separation of military and political power. In the emerging democracies of Eastern Europe and Latin America, military legitimacy depends upon a new generation of leadership to reshape authoritarian concepts of military professionalism and improve historically poor civil-military relations.[7]

The challenge of promoting the Constitutional values of democracy, human rights and the rule of law in emerging democracies necessitates leadership that understands the importance of civil-military relations to military legitimacy. Leadership must be provided in civilian as well as military environments, which requires balancing the requirements of warfighting with a professional style that promotes civil-military relations.

THE DIPLOMAT WARRIOR

The requirements of military legitimacy in the new strategic environment call for a new style of leadership that depends more upon knowledge and the power of persuasion than on command authority. Leaders in operations other than war must be able to motivate others, both military and civilian, without arousing hostility – a Webster's Dictionary definition of diplomacy. Diplomacy is out of place in combat, where success is synonymous with hostility. But proficiency as a combat leader is not a sufficient qualification for leadership in operations other than war.

A former military advisor to Saudi Arabia has compared the contrasting leadership skills required in peace and war, and defined diplomatic leadership as leading from behind. General William H. Riley, Jr. noted that:

> Leadership is defined as making it happen. Obviously, an aggressive, confrontational, results-at-any-cost mentality would be counterproductive with our Saudi counterparts. A General Patton would

probably be a miserable failure in developing rapport and achieving progress with the Saudis. A take-the-hill kind of attitude that attempts to tally quick results will not work well in the Saudi environment. We should lead from behind and encourage our Saudi counterparts to take the prominent role in planning, co-ordinating, directing, and controlling their projects.[8]

While the diplomatic style of leadership required of the military advisor contrasts sharply with the leadership traits required in combat, the two are not mutually exclusive; both styles of leadership are right (and legitimate) for their respective environments. Many combat leaders are versatile enough to be both great warriors and diplomats, but some are not; others, like Colonel David H. Hackworth, do not care to be diplomats and should not be put in a position to jeopardize military legitimacy.

Colonel Hackworth, the most highly decorated combat soldier alive, is the quintessential undiplomatic warrior. His heroic but unabashedly brash military exploits have been chronicled in his book *About Face*. Colonel Hackworth recalled an earlier effort by the Army (circa 1954) to develop a

... new breed [that was] kind of a warrior-diplomat; as bloodless ballistics seemed to be phasing out the role of fighters on future battlegrounds, the emphasis increased on the diplomatic side of soldiering.[9]

Colonel Hackworth and his rowdy warriors wanted no part of military diplomacy. When Hackworth served as a MACV advisor in Vietnam he was openly cynical of his ARVN counterparts. His leadership style was direct and forceful, if not intimidating; and he had little use for the finesse required in diplomacy. In fairness to Colonel Hackworth, however, there was little comparison between the environment in Vietnam after 1968 and that of Saudi Arabia in the 1980s.

One officer who exemplified the qualities of the diplomat warrior in an advisory role was never a combat leader. Major General Edward Lansdale nevertheless had the confidence and respect of those he advised, including Ramon Magsaysay who was Defense Minister and later President of the Philippines. Lansdale helped Magsaysay successfully counter the Huk insurgency in the Philippines during the 1950s. Their counter-insurgency operations reflected a commitment to democracy, human rights and the rule of law that remains relevant today. Lansdale's successful counter-insurgency philosophy was based on a sensitivity to the human, or social, dimension of conflict:

Lansdale had real concern for insurgents and a great deal of sympathy for their goals. He was at one with his old friend, Magsaysay, who once said, 'When a man is prepared to give up his life to overthrow his own government, he must first have suffered greatly.' Lansdale was likewise in agreement with Magsaysay's position that 'those who have less in life must have more in law'.[10]

Lansdale's moral principles were the foundation of his concept of military legitimacy, and they exemplified those of the diplomat warrior. He considered any deed 'which makes the soldier a brother of the people, as well as their protector', a worthy one. And he cited the ancient Chinese general, Sun Tzu, who considered military humanitarian assistance as 'moral law'. It was Sun Tzu who observed that 'to fight and conquer is not supreme excellence; supreme excellence is to conquer without fighting'. Lansdale also cited a later Chinese leader, Mao Tse-tung, who required his soldiers to act in accordance with orders, not to take anything from the people, and not to allow private interests to injure public interests. Lansdale used these principles to illustrate his belief that in political warfare 'the paramount object was to gain the loyalty of people who inhabit the land'.[11]

Lansdale combined the best qualities of the military ethic and professionalism with a respect for divergent views and a strong belief in individual rights and responsibility as a measure of integrity. He was equally at home in a military or civilian environment and was outspoken in his criticism of narrow-minded military leaders. Criticizing the emphasis on body count as a criterion for mission success and insensitivity to collateral damage in Vietnam, Lansdale challenged myopic military leaders to look at the moral dimension of their actions:

> True Americans, Lansdale warned, would avoid such actions. 'Open your eyes where you serve', he ordered. Be good soldiers. Win over local populations. See that troops behave with true military courtesy. Keep always a high code of honor. Prize integrity. Accord others the dignity that is their birthright. Act as a friend. Have empathy and humility. Offer a smile and a greeting in the language of the host country. Practise what you preach. Only those who act in such ways are true Americans. This strength we must have, or all else we possess and do will be without lasting meaning.[12]

Lansdale applied the above qualities in operations other than war before they were known as such. Their success demonstrated how the values of the Army ethic – duty, loyalty, integrity, selfless service and even compassion – can help make the military a positive and constructive force in achieving US foreign policy objectives during peacetime and conflict. When applied to civil-military relations, these traditional military values provide a context for ethical decision-making.

Vietnam was a crucible for the diplomat warrior, and until 1965 the Special Forces advisor reflected the Lansdale ideal. As late as 1969 there were diplomat warriors in remote hamlets of Vietnam still trying to salvage military and political legitimacy. One of these diplomat warriors wrote of his experiences in *Once a Warrior King*. He was invited back to Fort Bragg, where he once received training as a military advisor, to reflect on his experiences:

> I have seen the term in some of the materials here, the 'Diplomatic Warrior', and I suppose the things we did might be covered by such a term. On the civilian side, we worked for the American embassy on rural development, and on the military side we were tactical advisors for MACV. Almost every day and night we conducted military operations; two of us on the team were almost always away on an ambush or daylight operation. During the day we also provided advice and guidance for the civilian leaders.

The former 'warrior king' provided some timely advice to the civil affairs audience:

> I would say, yes, weave yourself into the local society if it helps the mission, but do not become lost in it. That is a danger. Remember who you are, where you are going, and why you are going there.

He emphasized that modern diplomat warriors should know the language, the culture, the objective, the resources available, and respect the people in their area of operations; but he noted that a belief in the democratic ideal is also required to sustain diplomat warriors:

> ... out there in those remote, unheard of villages around the world, in those lonely nights when no one seems to care, in those difficult times when the bullets fly and the bombs explode, in those days when the heat and bugs and the inefficiency of it all seem almost to have the victory; in those times that spirit within must still be able to guide you and to help you guide others. You and those you have been sent to help must be able to see out there in front of you the gleam of freedom rising, the faint flicker of justice awaiting. If you can help lead a people, even in small steps, toward those objectives you will have served your country and humanity well.[13]

The lessons of the former 'warrior king' have been as relevant in Haiti as they were in Vietnam. One Special Forces commander was described in 1994 as 'the contemporary version of a Roman procurator, the sole authority over the lives of three hundred thousand people living under primitive conditions in a mountainous, isolated four-hundred-square-kilometer administrative district'. In spite of many frustrations, Special Forces troops in Haiti have exemplified the spirit of the diplomat warrior:

> They know why they're in Haiti, even if the folks back home don't. Without guile or quixotic naivete, and with a growing, if not yet full, appreciation for the political and moral ambiguities of the mission, they say it: 'We're here to free the oppressed.' *De Oppresso Liber* is the motto of the Special Forces.[14]

That motto exemplifies military legitimacy in operations other than war, and gives meaning to the national values of democracy, human rights, and the rule of law. But as a leadership model the Special Forces soldier has been the exception rather than the rule. During the Cold War most civil-military operations overseas were considered special operations in LIC, the domain of Special Forces, Civil Affairs and Psychological Operations. While new Army doctrine incorporates peacetime civil-military activities into conventional operations other than war, the Army leadership model and the paradigm of the soldier and the state have not kept pace with that doctrine.

THE SOLDIER AND THE STATE

The traditional paradigm: the unpolitical soldier

The traditional paradigm of the soldier and the state and related concepts of military leadership and professionalism are based on a pure warrior ethic that calls for an unpolitical soldier, one clearly unsuited for civil-military operations in which political objectives predominate. If military leadership is to be reconciled with the requirements of legitimacy in operations other than war, this paradigm must be changed.

The author of the traditional paradigm is Samuel P. Huntington, the venerable Harvard professor who has described the current strategic environment as a clash of civilizations – one that requires coordinated military, economic, political and informational activities to achieve US security objectives. In such an operational environment military leaders must be able to bridge the formidable gap between military and political activities.

While Huntington's latest work implies the need for diplomat warriors in civil-military operations, his 1957 classic on *The Soldier and the State* minimized the importance of civil-military relations to military legitimacy, and described the military ethic as 'basically corporative [collective] in spirit and fundamentally anti-individualistic'.[15]

Huntington argued that civil-military relations should be minimal to avoid polluting the warrior spirit. He was convinced that military professionalism depended upon military officers remaining isolated from the politics of the civilian society they served. For Huntington, professionalism was defined by duty and loyalty to a military ideal: a robotic officer sworn to mindlessly obey hierarchical military authority, rather than a politically responsible officer who understood the principles and values of the Constitution and civil-military relations.[16]

While advocating the segregation of military and civilian activities, Huntington recognized the danger of conflicting military and civilian values to military legitimacy – or, as he termed it – the equilibrium of objective civilian control. For Huntington the unavoidable conflicts between an isolated military and civilian values were a price the military must pay to maintain the purity of its warrior ideal. His hope was that the lack of civil-military relations would be compensated by a shift in civilian support for his military ethic.[17]

A new paradigm: the political soldier

If there was ever a trend of civilian values toward Huntington's Cold War military ethic, it has since been reversed. The end of the 'evil empire' and defense budget constraints have upset the old equilibrium. Change is certain, but if contemporary missions are any indication, change will continue to be in the direction of developing more civil-military capabilities such as those of CA to balance traditional combat capabilities. This trend is reflected in the new Army doctrine on operations other than war.

Huntington was right about the need for an equilibrium between military and civilian values to ensure military effectiveness and legitimacy, but wrong in his predictions about changing civilian perceptions of military legitimacy. To achieve the equilibrium necessary to accommodate both military power and legitimacy in the new strategic environment, military values and concepts of professionalism must accommodate changing and sometimes conflicting civilian attitudes and values. The leadership paradigm of the diplomat warrior reflects a healthy balance of military and civilian values, a prerequisite for mission success in operations other than war.

The diplomat warrior is a political soldier who must understand the predominance of political objectives and the need for public support to achieve them. The diplomat warrior reflects the importance of civil-military relations to military legitimacy in operations other than war, where mission success requires that military leaders are knowledgeable in political affairs and work closely with civilians and foreign military personnel. An unlikely advocate, German General Ludwig Beck, warned of the dangers of military leaders isolated from politics, even in wartime:

'He who follows a false tradition of the unpolitical soldier and restricts himself to his military craft neglects an essential part of his sworn duty as a soldier in a democracy.' [Beck] warns that an officer corps that restricts itself to matters of craft may become indistinguishable from those Wehrmacht officers – honorable men by their own lights – who in doing their duty to the very end only propelled Germany that much further into the darkness. And he challenges us to embrace a mature vision of professionalism, [in which soldiers] appreciate the role of politics broadly defined in motivating, defining and guiding any genuinely effective military policy.[18]

These views contrast sharply with those of traditionalists who hold fast to the Huntington paradigm of the pure warrior ethic. One traditionalist dramatized the danger of politicizing the military with a hypothetical military takeover, led by military zealots who had acquired an insatiable appetite for political power through extensive civil-military activities. The protagonist, a colonel who resisted the coup, warned his friends of the dangers of politicizing the military: 'Demand that the armed forces focus exclusively on indisputably military duties.' The heroic colonel was in the mold of the pure warrior, isolated from the corrupting influence of civilian politics and values; he fit the Huntington paradigm of the soldier and the state.[19]

Another traditionalist was less subtle in warning that deteriorating civilian values could corrupt the pure values of military professionalism. He saw civilian values as a direct threat to military values and issued a call to arms:

The societal trends indicate a fundamental change in national values. The country's primary value-influencing institutions are promoting altered values for future recruits. These altered values are significantly different than the Army's values. The Army must preserve its integrity as an institution by resisting any tendency to accommodate these changed values.[20]

If there is a danger to democracy, it will not come from a military integrated with the society it must serve, but from an isolated military élite. An isolated warrior class is likely to develop authoritarian values that conflict with the libertarian values of the society it must serve. If there were to be a military coup, it would likely be to conform society to the authoritarian military ideal.[21]

But the remote threat of a military takeover is not the reason to change the traditional paradigm; it is the need for diplomat warriors in operations other than war. The unpolitical soldier – the pure warrior – cannot fulfil the requirements for leadership in operations other than war.

TWO SCHOOLS OF THOUGHT ON LEADERSHIP

The two paradigms of the soldier and the state reflect competing models of leadership which must be reconciled in the diplomat warrior. Nowhere is the traditional model of leadership held in higher esteem than at The Citadel, the Military College of South Carolina. Since 1842 it has made leadership its hallmark, with the 'Citadel Man' exemplifying the ideal citizen-soldier. That is, until 1993, when a female applied for admission to the all-male corps of cadets. In defending its single-gender tradition in the litigation that followed, the Citadel recommended a separate but equal leadership program for women, the Women's Leadership Institute (WLI).[22]

Columbia College was one of two women's colleges in South Carolina named to co-sponsor WLI. But the dean of the Leadership Institution at Columbia College resisted participation in WLI on the grounds that its traditional military model of leadership was incompatible with the model taught at her college:

> The central tenets of military leadership are conforming to clear directives from an officer, imitating the actions of a superior, standardization and regimentation, a 'win-lose' operating mentality and unquestioned allegiance to the chain of command.
>
> Leadership education at Columbia College has a different philosophical base. It is not hierarchical, nor does it focus on regimentation or repetitive drill. Hallmarks of this model of leadership are entrenched in the operating principles of collaboration, shared governance, commitment to seek 'win-win' solutions and decisions based on solid ethical premises.[23]

A retired Army general took exception to the above description of military leadership:

> After reading Dr. Mary Frame's explanation of military leadership, I was not sure what Army I served in for many years. She has a correct description of the old Soviet military leadership methods – always considered a weakness by Western military analysts.[24]

114

The Army War College teaches a situational approach to leadership that includes both the directive style of leadership needed in combat as well as more supportive styles required for operations other than war. Successful leadership in diverse operational environments requires a mix of both styles; there is no one best style of leadership for war and peace.[25] The Army, unlike The Citadel and Columbia College, is not a single-gender institution and cannot afford one-dimensional leadership. Its leaders must be flexible, equally at home in civilian and military environments and capable of employing both directive and supportive styles of leadership, depending on the situation.

The Army's model of leadership is incorporated in the concept of professionalism, which is expressed in the values of duty, loyalty, integrity and selfless service, and reflected in civil-military relations. As discussed earlier in this chapter, Huntington's classic on the soldier and the state described military professionalism from the perspective of the traditional paradigm, with its directive style of leadership, authoritarian concepts of duty and loyalty, and isolation from civilian politics. The new paradigm of the political soldier has its focus on the Constitution and incorporates the more supportive traits required in operations other than war, such as negotiation and diplomacy. Most of all, it encourages interaction between the military and the civilian society it serves to ensure healthy civil-military relations.

One of the best arguments for changing the old paradigm comes from its author, Samuel Huntington, whose description of the new strategic environment as one of clashing cultures makes his own traditional style of military leadership an anachronism, except in warfighting. For the military to be an effective instrument of national power in operations other than war it must have leaders whose concept of professionalism – their understanding of duty and loyalty – is consistent with the requirements and principles of military legitimacy.

LEGITIMACY, LEADERSHIP AND MILITARY PROFESSIONALISM

The military has traditionally considered itself a profession – the profession of arms. When this premise is challenged it can produce strong reactions from military officers. Is the Military Profession Legitimate? Colonel Lloyd J. Matthews answers his own question with an emphatic, 'Of course'. He acknowledges the lack of public support for the dirty business of warfare, berates writers who share negative public attitudes towards the military, and then states his lofty proposition

... that the military profession is the most vital, the most worthy of exaltation – and yes, the most legitimate – of all the professions ...[26]

To Colonel Matthews professionalism is vital to national security:

> If the nation's defenders are not members of a true higher calling and if that calling is not accorded the reverence of taxpayers and political leaders alike, then as surely as night follows day, the soldier's advice will come to be depreciated, the fighting forces and their leadership will be depleted of numbers and quality, and the security of this nation will fall into jeopardy.[27]

Colonel Matthews confuses professionalism with legitimacy. He apparently believes public acceptance of the military as a profession would ensure its legitimacy, generating public respect, even reverence for the military. Even in this he admits to having a long way to go, citing Samuel Huntington:

> ... the public, as well as the scholar, hardly conceives of the officer in the same way that it does the lawyer or doctor, and it certainly does not accord to the officer the deference which it gives to the civilian professions.[28]

The public hardly feels reverence and deference for lawyers, who are the butt of many jokes; but no one would deny their professional status. This illustrates the disconnection between professionalism and legitimacy overlooked by Matthews. Professional status does not necessarily bring public respect; in fact, it creates additional standards for the public trust and confidence needed for legitimacy – professional standards of conduct that will be discussed below.

Like his patron saint Samuel Huntington, Colonel Matthews justifies military professionalism (and legitimacy) exclusively on warfighting capabilities; but he does not make restraint in exercising those capabilities an issue of legitimacy. While he correctly defends the military against critics who have blamed them for the strategic errors of civilian policymakers, he fails to acknowledge restraint at operational and tactical levels as a primary issue of military professionalism (or legitimacy).

Rather than recognize that military legitimacy (a.k.a professionalism) is based on performance and recommend performance requirements, Matthews has chosen to discredit critics who have noted that military values can be at variance with those of the civilian society it must serve:

> There are some who would contest that the armed services of the state are properly professional on the grounds that ... military values are the antithesis of liberalism, the political philosophy upon which Western concepts of professionalism are based.[29]

The professionalism advocated by Colonel Matthews makes the requirements and principles of military legitimacy irrelevant. According to his arguments, when the inevitability of war and the need for a military is understood and the technical requirements of a profession are met, the military is *ipso facto* legitimate, without regard to its conduct. And since the inevitability of war justifies a military profession, its legitimacy is entirely dependent on its warfighting capability – never mind that it spends far more time with operations other than war than with warfighting. This logic makes any distinction between legitimacy in war and operations other than war irrelevant to issues of professionalism.

According to Matthews military professionalism depends upon proficiency in the 'unique defining specialty' of warfighting, which he implies would be compromised by the broader leadership responsibilities required in operations other than war. Matthews acknowledges 'the versatility and adaptability of the soldier in performing a broad spectrum of non-martial functions useful to the nation', but criticizes those who promote the diplomat warrior as a professional ideal:

> Additional to his sometime role as warfighter, in their view, he is diplomat, peacemaker, nation-builder, bureaucrat, advisor, teacher, manager, rescuer of the hurricane-beset, feeder of the starving and so on. I believe this claim is dangerously misleading.[30]

This narrow, traditionalist view of military professionalism does not do justice to the many senior officers who understand the importance of operations other than war and the capability of the military to perform them. Such a restricted view is also dangerous to the legitimacy of the military. Budgetary constraints and public attitudes preclude the maintenance of a large full-time military profession dedicated solely to warfighting. While warfighting must remain the priority skill of the military, its primary focus today – in assigned missions if not in future orientation – is not warfighting but managing violence in operations other than war. Contrary to the traditionalists, these are not mutually exclusive missions.

Military legitimacy is the primary component of military professionalism and, as discussed earlier, the standards are different for war and peace. In the absence of a pervasive and palpable threat to US security, military careerists had better find a way to reconcile their notions of professionalism with the versatility required in operations other than war – that versatility reflected in the diplomat warrior and civil affairs. A military profession that can only conduct combat operations is likely to become an anachronism in the new millennium.

117

MILITARY STANDARDS OF CONDUCT
AND PROFESSIONALISM

On one issue there is agreement: the need for a military code of professional conduct to consolidate the unique standards of conduct required for military legitimacy. In his list of elements for an ethical canon Colonel Matthews has included most of the requirements of military legitimacy: the oath of office (which incorporates the Constitution, democracy, human rights and the rule of law), military values, the Uniform Code of Military Justice (UCMJ), the laws of war and the Code of Conduct.[31] The only omission is domestic US law (OPLAW), which is discussed as it relates to specific military activities in Chapter 2. A code of professional conduct need not include all laws to which military personnel are subject – only those standards that are unique to the profession.

Every true profession has a code of professional conduct to hold its members accountable. But while the medical and legal professions have codes of ethics that define minimum standards of conduct expected of their members,[32] the military has no comparable code. The Army does have a 'Professional Ethic', but that ethic does not provide specific guidelines for ethical conduct at all, only institutional values.[33]

Why is there no military code of professional conduct? Defense doctrine on the subject reflects the views of military ethicists who think of ethics only in terms of values, and argue that specific standards degrade ethical values. Their emphasis on values is correct, but misplaced.[34] Without standards, values do not relate to behavior in a meaningful way, and there can be no accountability. And without accountability, professionalism has no real meaning.

The values of the Professional Army Ethic – duty, loyalty, integrity and selfless service – are the context of military legitimacy, but without standards they do not constitute professional ethics. If the military is to be considered a true profession, then like other professions it must identify its ethical standards of conduct and hold its members accountable to them.

Professionalism is defined as the conduct, aims or qualities that characterize or mark a profession.[35] The standards of conduct of the profession of arms, as with other professions, are those minimum standards required to perform its public trust. The public looks to military authority to regulate the profession according to its own standards, so long as they are consistent with the rule of law and national values. Military professionalism is inseparable from military legitimacy, and both reflect the quality of military leadership.

118

Military standards for the use of force

The management of lethal force requires special standards similar to those expected of civilian law enforcement officers. As with police officers, public confidence in the military is essential to its legitimacy; and ethical conduct is essential for that public confidence. For the military and police, the excessive use of force that violates human rights – military or police brutality – undermines public confidence and legitimacy.

Rules of engagement (ROE) provide the military standards that limit the use of lethal force and protect human rights. Peacetime ROE are based on self-defense and are similar to standards used by civilian law enforcement officers. While the improper or excessive use of force can be a crime, as evidenced in the trials of Los Angeles police officers involved in the Rodney King incident, as a standard of conduct it is more often the basis for taking administrative action rather than criminal action against the offender.

Military standards for conflicts of interest

In addition to restrictions on the use of lethal force, military personnel are subject to strict standards in the handling of public property. The *Standards of Ethical Conduct for Employees of the Executive Branch* published in August 1992 apply the Code of Ethics for Government Service enacted by Congress in 1978 to military officers, replacing AR 600-20, *Standards of Conduct*. These updated ethical guidelines prohibit conflicts of interest between public service and private interests. They reaffirm that government service is a public trust, and the corollary that using public office for private gain undermines public confidence in the government and the armed forces.[36]

The Department of Defense put special emphasis on standards of conduct after the Packard Commission's report to the President in 1986.[37] That report underscored the need to restore integrity as well as effectiveness to the government procurement process. Unfortunately, many military officers, both active and retired, have used public office for private gain. In so doing, they have undermined public confidence in the procurement process and military legitimacy as well.

The Code of Conduct

The only *Code of Conduct* denominated as such, and the one referred to by Colonel Matthews, does not provide standards at all and is not applicable to operations other than war. It was promulgated after the

119

Korean War to provide moral guidelines – not enforceable standards – for American personnel during combat and when captured. Its six articles are related to more specific, obligatory standards of conduct found in the punitive articles of the UCMJ and the Geneva Conventions Relating to Prisoners of War.[38]

The articles of the *Code of Conduct* are more like a supplement to the values of the Professional Army Ethic for prisoners-of-war than standards of conduct. They do relate to military legitimacy, providing guidelines for US prisoners-of-war. But any punishment for misconduct must be under the punitive articles of the UCMJ. In short, the *Code of Conduct* has a useful but limited purpose; it is not a comprehensive code of military standards.

Military standards vs. military crimes

While there is no code of military standards of conduct there is no shortage of such disciplinary standards. They are found in service regulations and the punitive articles of the UCMJ. Military standards of conduct are referred to as military crimes, although many are not true crimes since they have no civilian criminal counterpart.

Examples of non-criminal standards of conduct in the punitive articles are Article 86 (absence without leave), Article 89 (disrespect toward a superior commissioned officer), Article 91 (insubordinate conduct), Article 133 (conduct unbecoming an officer and a gentleman), and Article 134 (all disorders and neglects to the prejudice of good order and discipline, or of a nature to bring discredit upon the armed forces). These are only some of the disciplinary standards treated as crimes. Confusing the two has complicated concepts of military professionalism and military legitimacy.

Perhaps the best example is Article 133: conduct unbecoming an officer and gentleman. This article includes conduct below the minimum standards expected of an officer. Examples of prohibited conduct are making a false official statement, dishonorable failure to pay debts, cheating in an exam, opening and reading a letter of another without authority, and using insulting or defamatory language to or about another officer.[39] While these offenses give meaning to the value of integrity, they are not true crimes.

Closely related is general Article 134, which prohibits 'disorders and neglects to the prejudice of good order and discipline in the armed forces', and 'conduct of a nature to bring discredit upon the armed forces'. Listed under Article 134 in the *Manual for Courts-Martial* (MCM) is a formidable list of military standards of conduct intermixed with traditional and sometimes anachronistic misdemeanors, including abusing a public animal, wrongful cohabitation, disloyal statements, drinking liquor with a prisoner and fraternization.[40]

Even now, such misconduct is routinely punished administratively or by non-judicial punishment rather than by court martial.[41] This is because smart commanders have discovered that administrative procedures are less complex but provide adequate means to maintain good order and discipline. They have figured out for themselves that many military crimes are actually military disciplinary standards, and that commanders can better exercise their disciplinary authority by using administrative and non-judicial measures than by resorting to courts-martial.

But expediency should not be the sole reason for handling disciplinary measures administratively rather than applying criminal law. While military policy should conform to what has been recognized as good practise, the requirements of professionalism should also be a factor; and disciplinary action (as opposed to criminal prosecution) can be used to promote military professional development.

In the military the standards of professional conduct must begin with disciplinary standards. Punitive articles that are actually disciplinary standards should be decriminalized in order to promote professionalism. When disciplinary infractions are punished as crimes rather than standards of conduct, punishment has little relevance to professional development. When soldiers understand how misconduct relates to professional development, however, the disciplinary process can enhance professionalism.

Military standards in service regulations

Punitive articles are not the only source of standards for professional conduct. In fact, most military standards of conduct are found in service regulations. But since Article 92 of the UCMJ makes the failure to obey any lawful regulation a military crime, the standards of punitive regulations can be treated as criminal standards.

In addition to providing specific standards of conduct, service regulations also define the relationship between discipline and command authority which is at the heart of a military code of conduct. Army command policy in AR 600-20 provides:

> Military discipline is founded upon self-discipline, respect for properly constituted authority, and the embracing of the Professional Army Ethic with its supporting values.[42]

Among the military standards of conduct defined in AR 600-20 is the prohibition of 'improper relationships among military personnel'.

Improper relationships include sexual harassment, which is closely related to fraternization, a separate military offense under Article 134 of the UCMJ.[43]

The Tailhook scandal illustrates the relevancy of these standards to military legitimacy and leadership. The failure of the Navy chain of command to take prompt disciplinary action against those involved in improper relationships (sexual harassment) at a Las Vegas convention was a failure of leadership that compromised military legitimacy; it resulted in the forced resignation of the Secretary of the Navy, his top military lawyer, and the inspector general. Had the chain of command taken prompt disciplinary action in the matter, the resulting fiasco could have been avoided.

The requirements for a security clearance should be considered *de facto* professional standards since all positions of responsibility require such a clearance. The requirements are comprehensive and categorized as follows: loyalty, foreign preference, security responsibility, criminal conduct, foreign connections and vulnerability to blackmail, financial matters, alcohol and drug abuse, refusal to answer and sexual misconduct, including homosexuality.[44]

Homosexuality is not just a security concern. Personnel regulations describe homosexuality as incompatible with military service and make it a basis for involuntary separation from the service.[45] New regulations have been developed, but recent judicial decisions requiring the reinstatement of avowed homosexuals have cast their legality in doubt. Standards based on attitude rather than conduct are not likely to pass the test of constitutionality:

> The problem, as Tailhook so clearly reveals, already exists; the fundamental issue in the short run will not be attitude, but behavior. The services will have to review policies on acceptable conduct, on and off duty. Research on maintaining cohesion without scapegoating homosexuals and treating women as sex objects will have to be undertaken.[46]

There is likely to be more lively debate over military standards of conduct before the conflicting issues of sexual freedom and military effectiveness are resolved by policymakers and the courts. Continuing conflict between military and civilian values is certain; the level of tension has been evident in violent incidents between military personnel and homosexuals. Such violence cannot be condoned if the public trust and confidence necessary for military legitimacy is to be maintained.

It will remain difficult for the military to adapt to changing standards, such as those regarding sex; but the long-term danger of conflicting values to military legitimacy outweighs the short-term problems of transition. Standards for induction and separation, improper relationships, and fraternization are among those that must be periodically reviewed and modified to reflect changing societal norms. This does not mean that military standards must be identical to civilian standards, but significant deviations must be continuously validated as essential to good order and discipline.[47]

Enforcement: command influence and military justice

Public controversy over sexual harassment and homosexuality in the military reflects changing and sometimes conflicting civilian and military concepts of justice. The term military justice has come to be associated with the enforcement of military standards of conduct, whether as criminal prosecution or the administrative enforcement of disciplinary standards. Drawing a distinction between judicial and administrative enforcement procedures is just as important to professionalism and legitimacy as the distinction between crimes and military standards of conduct.

Public perceptions of military justice are based on civilian standards, and the traditional differences between military and civilian justice have raised issues of legitimacy. As an authoritarian system that emphasizes discipline, the rules and enforcement procedures of military justice seem harsh compared with more permissive civilian standards. The contrast is reinforced by a separate system of justice; the military is the only public bureaucracy with its own criminal court: the court-martial.

While differences remain between civilian and military justice, the latter has come a long way since the American Revolution. Then summary procedures and capital punishment were the norm. During the Civil War it was not unusual for a field commander to summarily execute a soldier who refused to fight. Due process in the military was non-existent; punishment was a command prerogative and military justice an oxymoron.

Today most of the constitutional protections accorded a civilian accused are available to those in the military. These include the Fourth Amendment right to be free from unreasonable searches and seizures; the Fifth Amendment right to remain silent and to due process (no person shall be deprived of life, liberty, or property without due process of law); and the Sixth Amendment right to counsel and to a speedy trial.

Even with the same fundamental rights as their civilian counterparts, the perception remains that the military accused has been shortchanged. That is because command influence – the pervasive effect of military authority – continues to plague military justice. And as long as commanders are involved in military criminal prosecutions, a fair and impartial trial by civilian standards will be illusory.

Because the military must remain an authoritarian organization in a democratic society, there will continue to be distinctions between military and civilian concepts of justice. Military legitimacy requires that these distinctions be minimal and absolutely necessary. Otherwise the unique characteristics of the court-martial, especially the potential for command influence, make it a lightning rod for public criticism.[48] One military law scholar has acknowledged the vulnerability of military justice to public criticism by describing it as 'a legal system looking for respect'.[49]

When commanders influence criminal prosecutions it contaminates the constitutional standards of due process. Command influence need not be egregious to be unlawful. Too often commanders say or do something that may influence a member of the court-martial or a witness, denying an accused his or her right to a fair and impartial trial. Command influence is pervasive since the authoritarian military environment is oriented to command directives. The Court of Military Appeals has described command influence as the 'mortal enemy' of military justice since it '... tends to deprive service-members of their constitutional rights'.[50]

The UCMJ specifically prohibits unlawful command influence,[51] and the military justice system has evolved to limit the role of a commander in courts-martial. As a result the court-martial has gradually become the province of military lawyers and judges. But senior commanders (convening authorities) still have significant roles: they convene (create) the court-martial, refer cases to it, appoint the court members, and approve their findings. These quasi-judicial powers make the line between lawful and unlawful command influence fuzzy; and they make impartiality, at least by civilian standards, impossible. These remnants of command-controlled justice in the military have been recognized as an impediment to its legitimacy, but there is disagreement over how to remedy the problem.[52]

During wartime and times of involuntary service (the draft) there are justifications for more command influence in military justice. The overriding needs of good order and discipline require a degree of coercion not required in an all-volunteer peacetime force, and the court-martial is accepted as a necessary instrument of military discipline. Command influence and limited due process, like collateral damage, are tolerated as the natural byproduct of war.

In a peacetime all-volunteer military the standards of legitimacy are

124

different. There is no justification for command influence in criminal trials. Criminal sanctions are not required to maintain discipline; and when service-members are accused of serious crimes there are civilian criminal courts to try them. Commanders can maintain discipline with administrative and non-judicial procedures which do not require the complex standards of due process required in courts-martial.[53]

With a decreasing number of courts-martial in the all-volunteer peacetime Army[54] and increasing pressure to reduce defense costs by eliminating non-essential services, the peacetime court-martial should become the responsibility of those reserve components that must provide it in wartime. Commanders would have a choice: criminal cases could be prosecuted either by civilian prosecutors in civilian courts or, if courts-martial were deemed necessary, by reserve component lawyers and judges. In neither event would command influence be an issue.

Reservist lawyers and judges could handle the relatively few peacetime courts-martial on a part-time basis, representing the government or the accused as they would any other client. It would provide reservists with the best training possible for their wartime responsibilities, and result in significant savings in the process.

In addition to savings, military legitimacy would also benefit from such a restructuring of the court-martial. Reservist lawyers and judges who regularly practise in civilian criminal courts would bring to the military courtroom a healthy mix of civilian and military values, helping moderate public suspicions of a separate and less equal standard of military justice. They would help bring much-needed public respect to military justice.

While command influence should be removed from the courtroom, it should be restored to the disciplinary process. Military discipline in peacetime should rarely require criminal prosecution. For violation of military standards of conduct, as distinguished from criminal standards of conduct, there are ample administrative actions and non-judicial punishments available to the commander short of court-martial. These disciplinary measures do not prohibit command influence – they require it.[55]

A MILITARY CODE OF PROFESSIONAL CONDUCT

A new sense of professionalism could result if minimum standards of military professional conduct were identified and those standards incorporated into a comprehensive military code of professional conduct. It would include military standards relating to sexual harassment and homosexuality, among others. Such changing standards reflect the dynamic nature of civil-military relations and their impact on military legitimacy.

For the all-volunteer peacetime military, a code of conduct would be enforced administratively much like analogous codes of professional conduct for the legal and medical professions. For them the most serious punishment for professional dereliction is loss of professional standing, the equivalent of an administrative separation from military service. Short of separation these professions, like the military, have a variety of administrative sanctions, beginning with a simple reprimand.

But the analogy between the military and the legal and medical professions can be carried only so far. Members of the civilian professions are not duty-bound to obey the orders of their superiors and to risk their lives in the line of duty. These distinctions make the military unique and justify coercion to ensure compliance with lawful orders. While administrative and non-judicial measures are adequate to maintain good order and discipline in an all-volunteer peacetime military, they must be supplemented by the criminal sanctions of court-martial in war and periods of involuntary service.

The unique nature of the military profession requires a more flexible code of conduct than those of the civilian professions. There must be more severe punishments for wartime offenses; failing to obey a lawful order should be a disciplinary measure in peace but a military crime in war. The punitive articles decriminalized for the all-volunteer force would revert to crimes in the event of war or the reinstatement of the draft.

SUMMARY

The requirements and principles of operations other than war call for a new paradigm of military leadership to bridge the gap between the limits of diplomacy and combat: the diplomat warrior. As the personification of military legitimacy both at home and abroad, the diplomat warrior exemplifies the traits of leadership and professionalism needed in peacetime to complement those of the wartime combat leader.

The traditional paradigm of the soldier and the state, with its ideal of an élite officer corps separated from its political environment and exclusively concerned with warfighting, is ill-suited for the new strategic environment. In a world where better civil-military relations are essential to peace and political stability, the diplomat warrior should be the new paradigm for military leadership and professionalism.

Leadership and legitimacy are inextricably bound up with the concept of professionalism, and a military code of professional conduct would enhance legitimacy by defining the minimal standards expected of those in the military profession. There is no shortage of standards of conduct or enforcement procedures for such a code; only a need to codify and identify them as what they are: disciplinary, not criminal, standards and enforcement procedures. Lessons learned in legitimacy have confirmed the need for military leadership imbued with a keen sense of professionalism, especially in operations other than war.

NOTES

1. The diplomat warrior is described in Barnes, 'Military Legitimacy and the Diplomat Warrior', *Small Wars and Insurgencies* (Spring/Summer 1993), at pp. 16–19.
2. *Student Text, Military Law and Justice* (Required Readings in Military Science IV, Military Qualifications Standards I, Precommissioning Requirements, June 1992); joint proponents for the test publication are the Judge Advocate General's School in Charlottesville, VA, and the Department of Law at the US Military Academy.
3. See Peter Maslowski, 'Army Values and American Values', *Military Review* (April 1990), p. 10.
4. See John E. Shephard, 'Thomas Becket, Ollie North, and You: The Importance of an Ethical Command Climate', *Military Review* (May 1991), pp. 21, 26.
5. Anthony E. Hartle, 'The Ethical Odyssey of Oliver North', *Parameters* (Summer 1993), pp. 28, 32–33. The Iran-Contra affair is used as a case study on 'covert war in peacetime' in Chapter 10 of *National Security Law*, edited by Stephen Dycus et al. (Boston: Little, Brown and Company, 1990).
6. See Barnes, 'Military Legitimacy and the Diplomat Warrior', n. 1 supra, pp. 5, 8–11, 18–19.
7. As to Eastern Europe, see Jacob W. Kipp, 'Civil-Military Relations in Central and Eastern Europe', *Military Review* (December 1992), p. 27. As to Latin America, see Gabriel Marcella, 'The Latin American Military, Low Intensity Conflict, and Democracy', *Winning the Peace: The Strategic Implications of Military Civic Action*, edited by John W. DePauw and George A. Luz (Carlisle, PA: Strategic Studies Institute, US Army War College), Chapter 4.
8. William H. Riley, Jr., 'Challenges of a Military Advisor', *Military Review* (November 1988), p. 34. General Riley's article preceded *Desert Storm* so that General Schwarzkopf's style of leadership was not considered.
9. David H. Hackworth, *About Face* (New York: Simon and Schuster, 1989), p. 315.
10. Cecil B. Currey, 'Edward G. Landsdale: LIC and the Ugly American', *Military Review* (May 1988), p. 50.
11. Ibid. at pp. 50–52.
12. See references at n. 1 to Chapter 2.
13. David Donovan is a pseudonym used by Dr. Terry Turner, the author of *Once A Warrior King* (New York: Ballantine Books, 1985). Dr. Turner

spoke to CA personnel at the John F. Kennedy Center for Special Warfare and School at Fort Bragg, NC in July 1993.

14. Bob Shacochis, 'The Immaculate Invasion', *Harper's Magazine* (February 1995), pp. 44, 59.

15. Samuel P. Huntington, *The Soldier and the State* (Cambridge, MA: Belknap Press of Harvard University Press, 1957), pp. 64, 70–94.

16. Ibid. at p. 84.

17. Ibid. at pp. 94 and 457.

18. A. J. Bacevich, 'New Rules: Modern War and Military Professionalism', *Parameters* (December 1990), pp. 12, 18–19.

19. Charles J. Dunlap, Jr., 'The Origins of the American Military Coup of 2012', *Parameters* (Winter 1992–92), pp. 2, 14.

20. Robert L. Maginnis, 'A Chasm of Values', *Military Review* (February 1993), pp. 2–11. See also n. 6 to Chapter 3, supra.

21. The danger of an emerging warrior class is discussed by Ralph Peters, 'The New Warrior Class', *Parameters* (Summer 1994), p. 16. In *The Transformation of War* (New York: The Free Press, 1991), Martin van Creveld extols warriors as those who love to fight: 'they are only too happy to give up their nearest and dearest in favor of – war!' (p. 227).

22. *The Citadel Alumni News* (Fall 1994), pp. 10–11.

23. Mary J. Frame, Ph.D., 'Columbia College, The Citadel teach different styles of leadership', *The State*, 8 February 1995, p. A9.

24. Thomas D. Ayers, Lt. Gen., USA Retired, 'Letters to the Editor, *The State*, 17 February 1995, p. A14.

25. Kenneth H. Blanchard, *A Situational Approach to Managing People* (Blanchard Training and Development, Inc., Esonhito, CA, 1985), pp. 2–4.

26. Lloyd J. Matthews, 'Is the Military Profession Legitimate?', *Army* (January 1994), pp. 15, 17.

27. Ibid., p. 17.

28. Idem.

29. Ibid., p. 18.

30. Ibid., at p. 22. In the same issue of *Army* magazine another professional purist, Col (Ret.) Michael D. Mahler, warns officers against 'pondering the diplomatic, political and multilateral fringes of our business' and urges them to refocus on 'basic doctrine and combat skills', Col Mahler cites General MacArthur's 1962 address at West Point: 'Your mission remains fixed, determined, inviolable – it is to win our wars ... All other public purposes, all other public projects, all other public needs, great or small, will find others for their accomplishment; ... Let civilian voices argue the merits or demerits of our processes of government ... These great national problems are not for your professional participation or military solution.' Such a myopic perspective of the military, focusing on the invincibility of US combat force and ignoring the potential of operations short of war, would ultimately contribute to America's worst defeat.

31. Matthews, n. 26 supra, at p. 23. In a subsequent article, 'The Need for an Officer's Code of Professional Ethics', *Army* (March 1994), at p. 21. Col Matthews proposes a code of ethics which would include these components, but does not distinguish between criminal acts and military standards of conduct as suggested here. Moreover, he does not believe enforcement is necessary, so that his code – unlike those for other professions – would have no enforcement provisions to ensure accountability. And like most other

military ethicists (see n. 34, infra) Matthews limits his ethical guidelines to career officers, which ignores the practical reality that ethical issues and responsibilities are not limited to the active component officer corps.

32. See *Department of the Army Rules of Professional Conduct for Lawyers*, 9 April 1991; also DA PAM 27–26, Rules of Professional Conduct for Lawyers (31 December 1987), and article by Dennis F. Coupe, 'Commanders, Staff Judge Advocates, and the Army Client', *The Army Lawyer* (November 1989), p. 3. The military lawyer is a member of a profession within a profession. He or she is subject to legal ethical standards as well as those required in the military profession. The format of the *Army Rules of Professional Conduct for Lawyers* could be used for rules and enforcement procedures applicable to all military personnel.

33. *The Professional Army Ethic* (the values of duty, loyalty, integrity, and selfless service) is set forth in FM 100–1, *The Army* (May 1986), Chapter 4. See also n. 56 to Chapter 6, infra.

34. Colonels Malham M. Wakin and Anthony E. Hartle are pre-eminent authorities and prolific writers on the subject of military ethics. They treat military ethics as a branch of moral philosophy, emphasizing democratic and traditional military values, such as those of the Professional Army Ethic, but avoiding standards. They correctly note the limitations of specific standards and the importance of values to professionalism, but overlook the importance of enforceable ethical standards for professional accountability. See Wakin, *War, Morality, and the Military Profession* (Boulder, Colorado: Frederick A. Praeger, 1986), Chapters 14 and 15; and Hartle, *Moral Issues in Military Decision Making* (Lawrence, Kansas: University Press of Kansas), Chapter 4.

35. *The Merriam-Webster Dictionary* (New York: Simon and Schuster Inc., 1974).

36. 5 USC 7351, 7353 (1978); *Standards of Ethical Conduct For Employees of the Executive Branch*, including Part I of Executive Order 12674 and 5 CFR. Part 2635 Regulation, Prepared by the US Office of Government Ethics, Washington, DC, August, 1992, Part I. See also DA PAM 27–21, *Administrative and Civil Law Handbook*, 1992; Chapter 12 provides an up to date discussion of conflicts between public and private interests that affect service members.

37. *A Quest for Excellence*, Final report to the President by the President's Blue Ribbon Commission on Defense Management, David Packard, Chairman (June 1986), Chapter 4, pp. 90–101.

38. The six articles of the Code of Conduct and the punitive articles of the UCMJ that relate to them are as follows:
Article I: I am an American fighting in the forces which guard my country and our way of life. I am prepared to give my life in their defense. (No punitive articles.)
Article II: I will never surrender of my own free will. If in command I will never surrender those under my command while they still have the means to resist. (*Punitive Article 99*: cowardice, disobedience, neglect of duty, or intentional misconduct in the face of the enemy; *Punitive Article 100*: compelling or attempting to compel surrender.)
Article III: If I am captured I will continue to resist by all means available. I will make every effort to escape and aid others to escape. I will accept neither parole nor special favors from the enemy. (*Punitive Article 90*: willfully disobeying a senior commissioned officer; *Punitive Article 105*: accepting special favors without proper authority to the detriment of other prisoners-of-war, and maltreatment of prisoners-of-war while in authority.)

Article IV: If I become a prisoner-of-war I will keep faith with my fellow prisoners. I will give no information nor take part in any action which might be harmful to my comrades. If I am senior I will take command. If not, I will obey the lawful orders of those appointed over me, and will back them up in every way. (*Punitive Article 104*: giving aid [e.g., information] to the enemy; *Punitive Article 90*: see above.)

Article V: When questioned, should I become a prisoner of war, I am required to give name, rank, service number and date of birth. I will evade answering further questions to the utmost of my ability. I will make no oral or written statements disloyal to my country and its allies or harmful to their cause. (*Punitive Article 104*: see above.)

Article VI: I will never forget that I am an American, fighting for freedom, responsible for my actions, and dedicated to the principles which made my country free. I will trust in my God and in the United States of America. (No punitive articles.)

39. See *Manual for Courts-Martial, United States, 1984* (as amended), hereinafter MCM, Article 133, p. IV–108.
40. See MCM, Article 134, pp. IV–109 – IV–147.
41. Administrative non-punitive disciplinary measures and non-judicial punishment available to the commander to maintain good order and discipline without resorting to court-martial include counselling, reprimand, admonishment, extra duties, restriction, reduction in rank, forfeiture of pay and allowances, bar to re-enlistment, and ultimately separation from the service under other than honorable conditions. See FM 27–1, *Legal Guide for Commanders*, 1987, Chapter 1 (Military Justice), Section IV, for a discussion of non-judicial punishment; and Chapter 2 (Administrative Law), Sections I and II, for a discussion of administrative and non-punitive measures. DA Pam 27–21, *Administrative and Civil Law Handbook*, 1992, Chapter 6, provides more information on administrative personnel actions; AR 27–10, *Military Justice*, 1989, Chapter 3, and *The Manual for Courts-Martial* (MCM), 1984 (as amended), part V, provide more information on non-judicial punishment under Article 15, UCMJ. See William Hagan, 'The Officer Corps: Unduly Distant From Military Justice?' *Military Review* (April 1991), p. 51. Colonel Hagan notes the frustration of commanders with military justice becoming less of a command function and the resulting wider use of administrative disciplinary procedures. He argues that a lack of understanding (of the proper role of command influence) is a large part of the problem and that ROTC cadets and OCS officer candidates should receive military justice training commensurate with that provided at the United States Military Academy at West Point.
42. See AR 600–20, *Army Command Policy*, 1988, Chapter 4, pp. 9 *et seq*.
43. See Art House, 'The "F-Word": Fraternization', *Army Reserve Magazine*, Fourth issue of 1991, p. 22. Colonel House attempts to distinguish fraternization (the military crime under article 134, UCMJ) from improper relationships among military personnel (the military standard of conduct under AR 600–20), but succeeds only in demonstrating that no meaningful distinction exists. Major David A. Jonas provides a comprehensive review of fraternization as an evolving custom of the services and disciplinary standard and critiques the new DOD definition of fraternization in 'Fraternization: Time for a Rational Department of Defense Standard', *Military Law Review* (Winter 1992), pp. 37, 41.

44. AR 380–67, *Personnel Security Program* (9 September 1988), Appendix I, pp. 46 *et seq*. The relationship of personnel security issues to national security law is discussed in Chapter 13, 'National Security Law'.
45. AR 635–200, Chapter 15, provides that homosexuality is incompatible with military service and that soldiers who are homosexuals will be separated from the service. Under this regulation, the homosexuality of the soldier can be established by pre-service, prior service, or current service conduct or statements.
46. Richard H. Kohn, 'Women in Combat, Homosexuals in Uniform: The Challenge of Military Leadership', *Parameters* (Spring 1993), pp. 2–3.
47. Ibid., at p. 4; see n. 7 at Chapter 3 where the quote was used in connection with values. Senator Sam Nunn has explained the constitutional basis for Congress requiring military standards of conduct at variance with those of civilian society in 'The Fundamental Principles of the Supreme Court's Jurisprudence in Military Cases', *The Army Lawyer* (January 1995), p. 27.
48. The conflict between civilian and military perceptions of justice in the military, especially the pervasive influence of commanders in courts-martial, was the subject of an investigative report by Ed Timms and Steve McGonigle of *The Dallas Morning News*, later published in *The State* (Columbia, SC), 5 January 1992, at p. 2-D under the headline: 'A Case of Military Injustice'. The article was the culmination of a three month investigation that purportedly found '... attempts by military commanders to influence courts-martial – a crime if intentional. Yet no commander has ever been prosecuted for such a breach of law.' The article compared civilian criminal justice with military justice and cited arguments both condemning and praising military justice, but offered no new solutions. It reflects a general civilian distrust of a system of justice run by an authoritarian organization in a democratic society.
49. Professor David Schleuter, a noted authority on military justice, has commiserated with military lawyers in acknowledging that the military justice system suffers from '... a lack of respect for the system by the public and legal profession generally'. He attributed this lack of respect primarily to the role of the commander in the court-martial, especially in choosing the panel of officers (the equivalent of a jury), and predicted continuing command influence in military justice until commanders are eliminated from the process. See David A. Schleuter, 'The Twentieth Annual Kenneth J. Hodson Lecture: Military Justice for the 1990s – A Legal System Looking for Respect', *Military Law Review* (Summer 1991), pp. 2, 10–23.
50. *United States v. Thomas*, 22 M.J. 388, 393–394 (CMA 1986).
51. See Article 37(a), UCMJ, and Rule 104, MCM. For examples see n. 49 supra. A practical list of command influence do's and don'ts is provided in Vito A. Clementi, 'Command Influence and Military Justice', *Military Review* (April 1988), p. 65.
52. Professor Schleuter has noted that the Court of Military Appeal has characterized command influence as the 'mortal enemy' of military justice. See n. 49 supra.
53. The standards of procedural due process of law guaranteed by the Fifth Amendment to the US Constitution differ according to the severity of the action to be taken. The military standards of *judicial due process* are those required in criminal prosecutions in which liberty and even life are at risk, and are essentially the same as those required for a civilian criminal defendant. But since the most serious punishment resulting from the administrative

enforcement of military standards of conduct is separation from the service (loss of professional status), the standards for *administrative due process* are less than those for judicial due process, minimizing the need for formal hearings, lawyers and judges in the process. Command influence invariably contaminates judicial due process, but the commander has an important role in administrative personnel actions. See Chapter 13, *Administrative Due Process*, DA PAM 27–21, *Administrative and Civil Law Handbook*, 1992.

54. Courts-martial, including general (GCM), bad conduct discharge special (BCD/SPCM), and special (SPCM) courts-martial have all decreased in recent years, while there has been little decrease in Article 15 non-judicial punishment:

	GCM	BCD/SPCM	SPCM	Art. 15
FY 88	1,631	923	182	50,066
FY 93	915	327	45	44,207

From *The Army Lawyer* (October 1994), at p. 15.

55. See article by Colonel William Hagan supra, n. 41.

6

Lessons Learned in Legitimacy and Leadership

Rescue those being led away to death,
hold back those who are being dragged to the slaughter.

Proverbs 24:11

Like catching a stray dog by the tail,
so is interfering in the quarrels of others.

Proverbs 26:17

Lessons learned from past successes and mistakes, especially those involving difficult issues of military legitimacy, must be incorporated in new strategies and force structures for operations other than war. The following lessons learned validate the concepts and principles of the preceding chapters and provide the foundation for the recommendations in the next chapter. They support the unique leadership requirements exemplified in the diplomat warrior and civil affairs (CA) as the primary capability for civil-military operations.

MILITARY LEGITIMACY AND PUBLIC SUPPORT

Public support has proved to be both a requirement and a measure of military legitimacy from the jungles of Vietnam to the deserts of the Persian Gulf, even to the killing fields of Somalia. These contrasting conflicts had one strategic common denominator: the legitimacy of the US military in each conflict depended upon public support back home.

Colonel (Retired) Harry G. Summers Jr., attributed the US failure in Vietnam largely to the unwillingness of President Johnson to mobilize the reserves and the national will when he escalated the US commitment from nation assistance to war – a war the president thought would be low profile and fought by professionals and draftees.[1]

Almost 20 years later President Bush mobilized the reserves and went to Congress for a resolution of support before ordering the *Desert Storm* offensive. The support of the American people for the war effort was mobilized with the reserves and the Congressional resolution of January 1991, which authorized the president to use offensive force.

The loss of public support for US involvement in Somalia following the abortive raid in Mogadishu in October 1993 was reminiscent of Vietnam. That painful lesson in legitimacy will be discussed further under the principles of objective and restraint.

Public support in the area of operations has proven to be even more important in nation assistance than in combat. Normally that support is essential to the legitimacy of the host government, and requires that US efforts be low profile to be effective. General George A. Joulwan has emphasized the need for US forces in Latin America to take a subordinate role in nation assistance:

> US agencies cannot become the recognized or perceived authority within any host nation, otherwise the legitimacy of the host nation's government is undermined. For the same reasons, USSOUTHCOM [US Southern Command] assists with the professional development of a host nation's military force while taking care not to assume its role. The concern for legitimacy underscores USSOUTHCOM's assistance role in the theater.[2]

POLITICAL AND MILITARY OBJECTIVES

When political and military objectives have not been made clear, commanders have often chosen inappropriate means to achieve them which have compromised military legitimacy. Vietnam is the classic case, where the lack of clear objectives allowed the body count to replace local public support as the measure of success. The US involvement in Somalia provides a more recent and relevant example.

President Bush demonstrated an understanding of the strategic dimension of legitimacy when he initiated *Restore Hope* in December 1991. Largely because political and military objectives were reasonably clear and understood by all concerned, phase one of the Somalia humanitarian assistance mission ended on a successful note in May 1993.

In the second phase President Clinton had a more difficult time with the strategic requirements of legitimacy. There were no clear US political objectives and no defined end state for US and other UN forces in Somalia. After the conclusion of *Restore Hope* it was not clear whether the UN mission was nation assistance, peacekeeping or peace enforcement; there was clearly no government to support and no peace to keep.

The dictates of PDD 25 would not have allowed US combat forces in such an ambiguous situation, but in the midst of uncertainty they remained, until mission creep eroded restraint and US forces abandoned the defensive. It was reminiscent of the fateful transition period in Vietnam: there was no public debate and no stated change in strategic objectives. If strategic decisions were made, they were not publicized.

The legitimacy of the US presence in Somalia was lost following the abortive 1993 raid in Mogadishu when the American public witnessed the body of a US soldier being dragged through the streets of that city. The strategic error of the US (and the UN) was the failure to clarify political and military objectives, complicated by the lack of any end state for the military forces involved.

Without clear strategic guidance combat commanders can be expected to use overwhelming force to achieve tactical military objectives without regard for political objectives. In Somalia it resulted in the ignominious withdrawal of US forces in March 1994, with the remaining UN forces withdrawn a year later.

In Latin America the US has avoided the pitfalls of mission creep. General Joulwan has recognized the importance of the requirements of military legitimacy and described the strategic objective of USSOUTHCOM as a political one that fosters

> ... a community of free, stable and prosperous nations acting in concert with one another while representing the dignity and rights of the individual and adhering to the principle of sovereignty and international law.[3]

UNITY OF EFFORT AND INTERAGENCY OPERATIONS

Nation assistance operations are interagency by their nature. Whether emergency disaster relief, continuing humanitarian and security assistance, or post-conflict, the military is but one element in nation assistance, and it is often subordinate to US civilian agencies. Mission success requires unity of effort between all participants; diplomat warriors must *lead from behind* when working with US foreign service personnel and an indigenous population.

In *Just Cause* problems developed when military planners failed to co-ordinate their operations with State Department officials:

To preserve operational security, the US Ambassador, the State Department, and other agencies of the government were not included in the initial planning process. This lack of coordination caused problems when different approaches surfaced during implementation. Senior SOUTHCOM leadership and the supporting task force failed to devote sufficient attention to integrating the civil-military operation with the tactical concept they had developed.[4]

In *Desert Shield/Storm*, after early indifference by military planners to overall political objectives and State Department responsibility for civil-military issues, CA reservists helped bridge the gap between military and dipomatic officials:

> Despite opposition or indifference on the part of key players, the Kuwait Task Force [made up of senior CA officers] was established, and it worked closely with the Government of Kuwait to plan and prepare for Kuwait's liberation.[5]

The refugee control operation at Guantanimo Naval Base in 1992 (*GITMO*) and again in 1994 required co-ordination with numerous civilian agencies in the lead: the Immigration and Naturalization Service, the Community Relations Service, the United Nations High Commission for Refugees, and the International Organization of Migrants. In Somalia (*Restore Hope*), non-governmental organizations (relief agencies) had important and sometimes leading roles. They were the primary service providers, and political advisors were often requested to help resolve difficult civil-military issues. Commanders at all levels had to practise the art of diplomacy, working closely with civilians, relating military activities to political objectives.

Operations *GITMO* and *Restore Hope* both taught that in operations other than war, diplomat warriors were required to bridge the gap between the military and civilian agencies of government, as well as non-governmental organizations, and then relate to local civilians in an unfamiliar cultural environment, a mission that necessitated overcoming international barriers of language and custom.

General Joulwan has emphasized the importance of unity of effort in Latin America:

> Most often, other US agencies will have the lead in operations other than war and will be supported with US military resources. Such is the case with USSOUTHCOM and other agencies committed in Central and South America – we support US ambassadors and their country teams, which are in the lead.[6]

PERSEVERANCE VERSUS THE QUICK FIX

Nation assistance and peace operations require perseverance, but strikes and raids do not. When given a choice the American public and its elected representatives have indicated a preference for the combat quick fix rather than more extended non-combat operations. In spite of security policy (the Weinberger Doctrine and PDD 25) that discourages the use of combat forces, the law and public sentiment encourage quick and dirty solutions to security problems overseas – so long as they are effective and produce few US casualties.

The War Powers Act limits to 60 days any commitment of US forces '... where imminent involvement in hostilities is clearly indicated by the circumstances' without the approval of Congress.[7] It is easy to understand why presidents have opted for overwhelming combat force in Vietnam, Grenada and Panama rather than for some measured but sustained force that might have achieved the same objectives with less violence and expense. The same impatience with US military commitments helps explain the Mogadishu debacle.

The lesson learned is that for both legal and political reasons extended non-combat operations should be avoided where hostilities are imminent. PDD 25 has confirmed that preference into policy. But *Provide Comfort*, the humanitarian and security assistance operation initiated after *Desert Storm* to protect the Kurds in northern Iraq from a vengeful Saddam Hussein, is an exception that proves the rule. It has quietly (except for the tragic friendly fire accident in 1994) accomplished its objectives, demonstrating the importance of perseverance to military and political legitimacy in operations other than war.

The danger of impatience to military and political legitimacy is illustrated by the abortive Mogadishu raid in October 1993. Had US forces maintained a defensive posture in Somalia they might not have won, but neither would they have lost. The win/lose dichotomy is irrelevant to the political objectives of operations other than war so long as combat forces do not take the offensive; when applied by impatient commanders it runs counter to the principle of restraint, escalating the level of violence and undermining military legitimacy.[8]

Experience has taught that when perseverance is required for mission success, as in nation assistance and peace operations, the requirements of military legitimacy become more difficult, while compliance is even more critical to mission success. Avoiding mission creep, complying with the law and prevailing cultural norms, and maintaining public support over a long period of time require diplomat warriors of unusual patience, ability and understanding.

General Joulwan has confirmed the need for perseverance to achieve US security objectives in Latin America:

> Success in USSOUTHCOM's AOR in operations other than war will be measured by lasting improvements in host nation stability, prosperity and respect for individual rights rather than short-term military victories. Perseverance won the peace in El Salvador, and perseverance is required to sustain the peace. Changes come slowly. Bringing them about requires a long-term commitment to help, nurture and reinforce success, just as a commander would do on a conventional battlefield.[9]

Culture clash has proven to be a continuing threat to military legitimacy in extended military operations. In late 1990 during the *Desert Shield* build-up there were few local civilians in the desert to complicate issues of military legitimacy. Even so there were numerous culture clashes: Christian and Jewish religious services created a stir among Islamic Saudis; and recreational activities involving alcoholic beverages created even more serious problems. Finally, the relative equality and independence of women in the US military was too much for conservative Saudis to tolerate. Had *Desert Storm* not subordinated cultural concerns to the war effort and a quick victory not allowed most US forces to be withdrawn shortly thereafter, the legitimacy of the US presence would have been in jeopardy.

In *GITMO* and in Somalia there were cultural clashes along religious and racial lines, and language was often a barrier to resolving these conflicts. Behavioral standards in these unfamiliar cultural environments were not those to which US forces were accustomed, and civilian riots resulted when communication and security measures failed. Linguists were essential but rarely part of military units; and few military leaders had the language capability needed. Experience in both *GITMO* and Somalia confirmed that effective leadership requires a combination of perseverance, understanding, tolerance and firm action.

RESTRAINT IN THE USE OF FORCE

The danger of excessive force to military objectives is the most important of all lessons learned in legitimacy. Wherever excessive force has caused collateral damage or injury to innocent civilians the legitimacy of US military forces has been called into question. Experience has proved the relevance of the Just War principles of just cause and right intention at the strategic level, and discrimination and proportionality at the operational level.

Strategic restraint was a lesson learned in Vietnam. Painful experience there taught that the lack of restraint (or excessive force) can result in the loss of the public support required for military and political legitimacy, both at home and in the area of operations. A corollary lesson learned was that military victory achieved through superior military force can be irrelevant to political legitimacy.

> 'You know you never defeated us on the battlefield', said the American colonel. The North Vietnamese colonel pondered this remark a moment. 'That may be so', he replied, 'but it is also irrelevant'.[10]

The Weinberger Doctrine and PDD 25 confirm that combat force should be the military measure of last resort in peacetime, but that does not mean that *all* military capabilities should be a last resort in peacetime – that is, unless all military capabilities are combat forces.

While combat must remain the primary purpose of US military forces, it is not the only military capability. Combat support and service support forces provide the primary capabilities needed for non-combat operations other than war. If combat forces were the only military option there would be considerable danger that military force would be used inappropriately.

Non-combat operations other than war can extend US power beyond the limits of diplomacy but short of combat; they do not involve the US in a win-lose situation any more than other forms of foreign assistance. Operations other than war are not a substitute for wartime operations; they complement them. It is critical that strategists understand the distinction between the two and do not substitute one for the other.

Colonel Harry Summers has pointed out the danger of confusing operations other than war with warfighting. He argued that the 1968 version of FM 100-5 degraded US combat capabilities in Vietnam by confusing counter-insurgency and warfighting doctrine. Counter-insurgency was the Vietnam era equivalent of operations other than war, and represented a dramatic change in doctrine from the 1954 version of FM 100-5:

> This change ... stated that 'the fundamental purpose of US forces is to preserve, restore, or create an environment of order or stability within which the instrumentalities of government can function effectively under a code of laws'. We had come a long way from the pre-Vietnam war doctrine that called for 'the defeat of an enemy by

application of military power directly or indirectly against the armed forces which support his political structure'.[11]

Colonel Summers cited General Fred C. Weyand on the limitations of combat force to achieve political legitimacy:

> But there are fundamental limitations on American military power ... the Congress and the American people will not permit their military to take total control of another nation's political, economic, and social institutions in order to completely orchestrate the war ... The failure to communicate these capabilities and limitations resulted in the military being called upon to perform political, economic, and social tasks beyond its capability while at the same time it was limited in its authority to accomplish those military tasks of which it was capable.[12]

Though skeptical of counter-insurgency doctrine in the context of Vietnam, Colonel Summers grudgingly acknowledged that even in Vietnam nation assistance was an appropriate strategy until President Johnson escalated the US role to warfighting:

> Where did we go wrong? It can be argued that from the French withdrawal in 1954 until President Diem's assassination in 1963, the American response was essentially correct. The task at hand was one of assisting South Vietnam to become a viable nation state, and US military advisors contributed to that end.[13]

Colonel Summers and General Weyand may not have intended to support the strategic value of operations other than war, but they do make a good case for combat force being the military measure of last resort in peacetime. Neither argues against non-combat military capabilities; their point is that combat is the primary purpose of the armed forces and should not be denigrated by non-combat operations other than war.

Using their logic the US could and should have avoided the tragedy of Vietnam by reducing its commitments after 1963 rather than deploying combat forces. Until combat forces were introduced, Vietnam was not a win-lose proposition for the US; but when US combat forces became engaged it put America's prestige on the line, so that anything short of military victory was a defeat. Restraint at the strategic level requires that combat forces be the last resort; non-combat operations other than war can further US security objectives without the risk of a military defeat.

Issues of restraint and military legitimacy arose again following US interventions in Grenada (*Urgent Fury*) in 1983 and Panama (*Just*

Cause) in 1989, with contrasting results. In Grenada military force was restrained and collateral damages minimal; there was widespread public support for the intervention and long-term political objectives have since been achieved. The same cannot be said for Panama. The short but violent combat phase was perceived as a success in the US, but the success of the civil-military operation which followed, *Promote Liberty*, remains in doubt; the 1994 elections restored Noriega's political party to power.

One study concluded that a lack of restraint jeopardized political objectives and legitimacy in Panama:

> There is little dispute that General Thurman, CINCSOUTH, and his planners concentrated on winning the war and paid insufficient attention to what might be required to win the peace afterward. Destruction of the Panama Defense Force (PDF) was not an assigned mission for SOUTHCOM but was thought by General Thurman and his planners to be a necessary prerequisite for restoring Panamanian democracy. Emphasis on destroying the PDF appears to have distracted attention from the civil aspects of the mission, whose purposes were to replace a corrupt, unelected Panamanian government with a new, legitimate, democratic government.[14]

Many Panamanians apparently agreed that *Just Cause* was a misnomer, and that excessive force was used to achieve the stated political objectives. This was reflected in a 1990 poll published in the Panamanian newspaper *La Prenza* which indicated that most felt the problems caused by the intervention outweighed its benefits. The president of the Panamanian Bar Association, Jose Alberto Alverez, may have captured the public sentiment:

> Of Bush's objectives, only one was really achieved – getting rid of Manuel Noriega ... (and) they could've captured him without an invasion, without destroying the country.[15]

Desert Storm followed *Just Cause*, and was even more destructive. But there was a critical distinction between the two: in *Just Cause* the support of the Panamanians was needed to achieve US political objectives; in *Desert Storm* the support of Iraqi civilians who bore the brunt of the destruction was not a US political objective. The primary political objective in *Desert Storm* was to restore Kuwait to the Kuwaitis.

Desert Storm was a war, albeit a short one, with different standards of military legitimacy. Civilians in Baghdad were perceived as the enemy, little different from those in Germany and Japan in the Second World War. This distinction explains the contrasting standards of restraint and also explains why Saddam Hussein remains in power.

The abortive US raid in Mogadishu in October 1993 was the most recent error in strategic restraint. Remembering President Johnson, who vowed he would not be the first president to suffer defeat and continued to escalate US military force in Vietnam after 1965, President Clinton did not hesitate to cut his losses and bring the troops home. But because US combat forces had been engaged in offensive action it was a humiliating defeat for American prestige abroad.

Operational restraint involves the use of weaponry; its standards are reflected in ROE which restrict lethal force to the minimum required to achieve military and political objectives. For operations other than war the standard for restraint is normally self-defense. But even the most carefully drawn ROE cannot resolve all the ambiguities for combat forces in contemporary conflict when combatants cannot be distinguished from non-combatants.

This was illustrated by the frustrations of Lieutenant Caputo recounted in Chapter 1. The ambiguity of the Vietnam conflict made it difficult, if not impossible, to make a distinction between combatant and non-combatant – the distinction upon which the legitimate use of lethal force in wartime depends, and a distinction lacking in operations other than war. In the face of such ethical ambiguity the massacre of My Lai should have been no surprise; nor the erosion of public support required to sustain US involvement.

Operation *Restore Hope* conducted in Somalia from December 1992 to May 1993 was the next extended US operation involving combat forces. The Marine Corps and Army forces involved exercised commendable restraint in complying with ROE which limited force to self-defense, but there were several incidents of excessive force that underscored the danger of using combat forces in operations other than war.[16]

Operational restraint goes beyond ROE. Again, Vietnam provided timeless lessons learned in legitimacy: any hope of public support there was compromised by the abusive way American soldiers and their South Vietnamese counterparts treated the populace:

> ... [I]t was obvious to even the rosiest fantasts that we couldn't win this war by simple force of arms, that the real battle was for the trust and loyalty of the common man. We knew this, but our anger and fear kept getting the better of us. Why didn't they get behind us? Why didn't they care that we were dying for them? Yet every time we slapped someone around or trashed a village, or shouted curses from a jeep, we defined ourselves as the enemy and thereby handed more power and legitimacy to the people we had to beat.[17]

Restraint is even more important to the legitimacy of domestic military operations than those overseas, as evidenced by the tragic incident at Kent State University on 4 May 1970, and the public outrage that followed. For reasons still unclear (subsequent FBI investigations indicated the troops were in no danger), soldiers of the Ohio National Guard opened fire on students during an anti-war demonstration. Of thirteen students hit four were killed – one of them a Reserve Officer Training Corps (ROTC) cadet.

The lessons of Kent State were not lost on members of the California National Guard who were activated during the Los Angeles riots of 1992. Under stressful conditions they scrupulously complied with ROE which severely limited their use of lethal force. Of only 19 rounds fired during several weeks on duty, 18 were recovered. One person was killed, but clearly in self-defense.[18]

SECURITY THROUGH LAW AND ORDER

Law and order are the essence of security and a prerequisite of military and political legitimacy. Where civilian law enforcement agencies cannot provide security US forces must be prepared to enforce basic standards of justice, including the protection of human rights, through the rule of law. When providing law and order, the military must act with the same restraint as police officers. Excessive force undermines both military and political legitimacy.

The failure to plan for civilian security can jeopardize military legitimacy, as it did in *Just Cause*:

> One result of the general lack of attention to the civil aspects of the operation [*Just Cause*] was a breakdown of law and order in Panama in the course of the US military operation ... One of the basic functions of CA is law enforcement, and proper use of CA planning would have foreseen the problem and prepared actions to preclude the collapse of law and order in Panama.[19]

Somalia provided examples of how the lack of security can be as fatal to legitimacy as the lack of restraint. Feuding warlords kept UN forces barricaded and of little use as roving bandits terrorized the country. Among the lessons learned in Somalia is that peacekeeping cannot occur where there is no peace to keep, and that combat forces should not be deployed unless the standards of the Weinberger Doctrine and PDD 25 can be met.

In the Haitian refugee camp at Guantanimo Naval Base (*GITMO*) there were rules but no uniform enforcement procedures; and before

143

suitable enforcement procedures could be implemented several riots occurred. Control measures required personal identification with photo identification cards, and fingerprinting for those causing trouble. Weapons control required a separate permitting process. Reliable refugees were used as enforcement officers to the extent possible to minimize resistance to security measures.

The difficulty of maintaining a balance between restraint and security was evident in operation *Uphold Democracy* in Haiti. Special Forces troops were charged with providing a 'safe and secure environment' and worked closely with local public officials to provide a security system to fill the vacuum after General Cedras and his cronies were ousted. But unrealistic public expectations frustrated the troopers with incessant demands for security that could not be met. One exasperated team commander complained he was 'sick and tired of being in the (expletive) police biz, and I'm just about ready to let the Haitian soldiers loose and beat the (expletive) out of everybody like they used to do'.[20]

CIVIL-MILITARY RELATIONS AND
HUMAN RIGHTS OVERSEAS

Continuing nation assistance operations have confirmed the importance of human rights and civil-military relations to military legitimacy. Military forces in undemocratic regimes have historically been associated with human rights violations and political oppression. The US has provided security and humanitarian assistance to encourage democratization in many such regimes: first to Latin America in the 1970s and 1980s, and more recently to Eastern Europe and Russia.

In most developing countries the military is not separated from domestic politics as it is in the US, and is expected to provide internal as well as external security to the civilian population. This makes it especially important that the requirements of internal security be balanced with the restraint needed to protect human rights.

In Latin America the exercise of political power by military forces has had a corrosive influence on military professionalism and civil-military relations. The use of the military to protect or install corrupt regimes has eroded their legitimacy. To build the public support required for military legitimacy, Latin American militaries must limit their political role, constrain the use of force and focus on improving civil-military relations.[21] A joint project between US Army military lawyers and their Peruvian counterparts to promote human rights is discussed in Chapter 7; it illustrates how institutionalizing respect for human rights in the military can improve civil-military relations and legitimacy.

According to the 1993 UN Truth Commission Report on the civil war in El Salvador, the Salvadoran military has a long way to go to achieve legitimacy. The Report indicates that gross human rights violations were committed by death squads associated with US-trained military forces. The conviction of two officers for the murder of six Jesuit priests in 1989 was a positive step, but the failure to purge those officers identified with human rights violations shows that military legitimacy remains an elusive goal.

In Eastern Europe there have been similar problems with military legitimacy. Promoting the values of democracy, human rights and the rule of law in former communist countries can contribute to peace, security and military legitimacy in two ways: first, democratic regimes are less likely to misuse their military power than authoritarian regimes; and second, the democratic values of individual liberty and civilian control contribute to better civil-military relations and military professionalism, and professionalism (internal control) is the best defense against the misuse of military power.[22]

The failure of civil-military relations and military professionalism in the 'masterless' armies of the former Yugoslavia has contributed to the unspeakable atrocities in the Bosnian civil war. Better civil-military relations and military professionalism could help protect human rights and prevent the spread of similar violence throughout the region.

> Civil-military relations take on a deeper significance and must be viewed as a critical element in the struggle to maintain legitimacy of existing democratic governments as they attempt to deal with the internal and external manifestations of this crisis.[23]

The future of democracy, human rights and the rule of law in emerging democracies depends upon better civil-military relations, which will require effective separation of military and political power (ideally civilian control of the military), but this should not prevent domestic military missions with political implications. For US military advisors to help their indigenous counterparts improve civil-military relations, they must understand the requirements and principles of military legitimacy, especially the role of human rights and the rule of law.

CIVIL-MILITARY RELATIONS IN THE US

There have been important lessons learned in civil-military relations in the US as well as overseas. While they have not involved serious violations of human rights (with the exception of the Indian wars and the Kent State incident) they do provide important lessons in military legitimacy and useful precedents for the future.

The US Army has been closely integrated with civilian society over its 219 years of existence, often as a matter of military and political necessity. Charles Heller has noted that the Army has spent far more time with domestic civil-military activities than with war-fighting:

> The Federal Army has: explored and mapped the continent, pioneered medical science, forecast the weather, delivered the mail, quelled civil disturbances, provided disaster relief, run youth programs, engaged in public works projects, and a host of other non-military missions.[24]

In *A Time to Build*, Henry F. Walterhouse noted many examples of successful civil-military activities that have contemporary applications.[25] His first example of civil-military co-operation came from West Point – an appropriate place to begin since Samuel Huntington used West Point to symbolize the traditional military ethic discussed in the previous chapter.

The Army Corps of Engineers

In reverent terms, Huntington compared the 'ordered serenity' of West Point to the 'commonplace' Highland Falls, a village just outside its gates.[26] Huntington's idealized comparison of his West Point Camelot and its inferior surroundings illustrates a subtle but important conflict between military and civilian values. The commonplace diversity of Highland Falls represents the ambience of a free society, while the structured uniformity that characterizes West Point is the insipid norm for authoritarian regimes.

Huntington's West Point could not have been imagined by George Washington and his fledgling Corps of Engineers when they made it their home in 1794. It did not become the US Military Academy until 1802, and for many years was the nation's only engineering college, scarcely distinguished from its civilian surroundings. The demands of an undeveloped frontier placed a high priority on engineering skills, and West Point and its Army Corps of Engineers met the challenge for civilian as well as military needs. Its faculty was also the primary source of academicians to staff the new technical schools developing across the country.[27]

Perhaps the Army's greatest civil-military project was the Panama Canal, completed in 1914 after eight grueling years battling the elements and tropical diseases. The engineers shared the victory with

Army doctors, who were instrumental in winning the battle against malaria and yellow fever in the festering Canal Zone. When President Teddy Roosevelt turned to the Army and chose Goethals to direct the project after earlier failures, he remarked:

> '... the great thing about an Army officer is that he does what you tell him to do'. During World War II, McGeorge Bundy recalled these words when Teddy's cousin, F.D.R., was considering the Army for military government in Europe. Bundy noted: 'Discipline without brains was of little value, but both Roosevelts learned to their cost the uselessness in administration of brains without discipline.'[28]

Today the Army Corps of Engineers continues its civil engineering projects and manages national flood control. It plays a major role in disaster relief, and has the potential for managing a wide range of environmental projects. With offices in major cities around the country, it provides a model for civil-military co-operation. For the Corps, civil engineering has a double meaning that requires its officers to combine public service with military skills – the attributes of the diplomat warrior.

The military in domestic civil administration

In spite of the traditional aversion of the military to domestic politics, there have been occasions when politics and the military have intentionally been mixed. Until 1849 the Army was responsible for Indian affairs – perhaps because the native Indian was their primary adversary. Among those officers assigned duty in Indian affairs was Lieutenant Colonel Zachary Taylor, who administered an Indian school for two years. Colonel Taylor, like most of his fellow officers, preferred combat command and was not happy with a civil-military administrative assignment, but it may have prepared him for his later political career.[29]

During the Mexican War, Army and Navy officers administered civil government in California, and continued until it became a state in 1850. Later the Army was called upon to administer the huge Alaska territory. Military government was required because conditions on the early frontiers were just one step removed from anarchy. There were problems, especially in California, related to jealousies and rivalries among military leaders who lacked political skills; other difficulties were inherent in the lawless and undeveloped environment. But the military demonstrated it could perform civil administation in primitive and violent conditions when required to do so.[30] The legacy was recalled when widespread lawlessness in the 1992 Los Angeles race riots required military assistance.

Following the American Civil War, the Army was called upon to administer the Freedman's Bureau, which was responsible for the welfare of former slaves. It was required to establish schools, distribute relief supplies, regulate labor, administer justice and provide medical assistance, job training and limited land redistribution. In many ways the responsibilities of the Freedman's Bureau were similar to those of modern nation assistance and refugee control.[31]

Civil-military youth programs

The Great Depression gave birth to the Civilian Conservation Corps (CCC), another domestic civil-military project relevant to contemporary public needs, both in the US and overseas. The CCC was formed in 1933 for the jobless, putting them to work on national resource and conservation projects. The no-frills program was successful because the military personnel who administered it understood the importance of civil-military relations.

When the CCC was initiated in 1933, the active Army totaled only 137,000 officers and men. But within the year 3,641 officers from the Regular Army, Navy and Marine Corps and 1,774 reserve officers were administering the camps. By 1935, enrollment in CCC camps exceeded half a million men, and more than 9,000 reserve officers had assumed the primary burden of camp administration. The CCC began to wane as the economy improved in 1939 and was terminated with the onset of war in 1942.[32]

In South Carolina alone, 200,000 erosion control dams, 97 fire towers, 25 parks and recreation areas and 57,000,000 trees were added to the landscape.[33] The Army provided something essential for the unemployed young people in the CCC: discipline. Those selected (less than half of the 8,000 applicants in South Carolina) were treated like military recruits – given medical examinations, issued uniforms and fed well.[34]

The accomplishments of the CCC in conserving both human and natural resources have inspired a new generation of civil-military programs. Senator Sam Nunn (D-Georgia) used the CCC as one of the models for the Civil-Military Co-operation Program which was incorporated in the 1993 defense bill. It authorized military assets to be used for such projects as infrastructure and housing construction and repair, and the education and training of disadvantaged youth.

Domestic civil-military activities can promote better civil-military relations through a sharing of values and provide opportunities and discipline for disillusioned young people. The military values of duty, loyalty, integrity and selfless service can inspire civilians as well as the military; and a new generation of American youth need an introduction to discipline. The CCC demonstrated how civil-military activities could contribute to military legitimacy half a century ago; the same concepts can be adapted to stem the troubling trends plaguing the US today.

The lessons of history are evident in a new generation of civil-military activities that have been emerging across the country. They relate the military to civilian needs in a positive and constructive way. Military-style boot camps for youthful offenders in California are an example:

> From personal observation and interviews with inmates and corrections officers, the results were remarkably successful. People on the lowest rungs of society can be salvaged by using military techniques and training to build self-respect, confidence and structure into their lives.[35]

Other new civil-military youth programs reach beyond youthful offenders. From a summer camp that teaches survival skills to under-privileged youth in San Antonio to a proposal for Special Operations Forces to provide military civic action in a Montana Indian reservation,[36] such programs serve public needs related to internal security, enhance military proficiency and improve the civil-military relations so essential to military legitimacy.

CIVIL AFFAIRS: FROM MILITARY GOVERNMENT TO NATION ASSISTANCE

Civil Affairs (CA) provides an operational concept and structure well suited for civil-military activities at home and abroad, as will be discussed in the next chapter. CA was born in the Second World War out of the need for military government in occupied territory. Since then CA personnel have provided an interface between the military and civilians and helped mobilize public (civilian) support when required for military and political objectives.

The Second World War illustrated how the role of CA changes dramatically when hostilities cease and the transition to peace begins. As the Allies drove into Germany the people who had been obstacles to combat operations became the objectives of peacetime military operations. In newly liberated areas the military was required to establish temporary governments to provide essential services to civilians. This was a command responsibility, and combat commanders were quick to recognize CA personnel as force multipliers, relieving their combat troops for battlefront duty.[37]

General Dwight D. Eisenhower initially requested 960 CA officers, and their number was later increased to thousands of personnel. Of these, approximately 200 were highly qualified lawyers, most of whom were assigned to military government duties. In some instances, however, unit staff judge advocates also provided legal support to US military governments.[38]

The emphasis on lawyers in military government reflected the priority of legitimacy in civil-military operations. The legal standards applicable to the treatment of civilians and the provision of essential legal services were priorities of military government. These priorities have since been incorporated into a principle applicable to all CA operations and activities: to ensure that commanders comply with their legal and moral obligations to civilians.[39]

During the Second World War, CA was synonymous with military government. Early post-war doctrine made a distinction between CA and military government by operational environment: operations in friendly countries were considered to be CA, while those in occupied enemy territory were military government. In and after the war, CA operations (now known as civil administration) were conducted in North Africa, France, Holland, Belgium and the Philippines, while military government operations were conducted in Sicily, Austria, Germany, Okinawa, Japan and Korea.[40]

Okinawa, the largest island in the Ryukyu chain southwest of Japan, illustrated how military government evolved into other forms of CA. The battle of Okinawa was the only US combat operation conducted against a Japanese land area with a large civilian population. The island was devastated by the fighting, with 95 per cent of its housing destroyed. With a population of shell-shocked, sick, homeless and scared people, Okinawa placed a tremendous welfare burden on the military.[41]

The Army responded to the challenge. Its military government provided tractors, farming implements, seeds, fertilizers and farm animals to restore agriculture; it helped restore the fishing industry with boat construction, provided essential utilities, and began long-range industrial promotion projects. The Army also promoted political and social reform; it helped to revise legal codes, re-established courts, introduced universal suffrage and promoted improved education.[42] In later years it conducted civic action projects throughout the Ryukyu Islands. By the time Okinawa reverted to Japan in 1971, it was a relatively prosperous island.

Korea represented another step in the evolution of CA from military government to nation assistance. Following the Second World War the first CA effort consisted of the Army disarming and repatriating Japanese troops, maintaining law and order, and providing essential public services. This phase of CA operations came to an end in August 1948,[43] but the new Korean administration lasted less than two years. The invasion by North Korea in June 1950 overran all of South Korea except the small port city of Pusan.

The role of CA was minimal during the Korean War, but afterward a variety of CA projects helped Koreans rebuild their country both politically and economically:

> Troop units down to company size, while maintaining their combat readiness, engaged in a comprehensive plan of repair, renovation, and restoration. Korean civic leaders indicated requirements; Korean agencies furnished local materials and labor, while US forces provided engineering skills, equipment, and the essential constituents for reconstruction. They concentrated efforts on schools, hospitals, civic buildings, land reclamation, and improvement of public health and transportation facilities.[44]

The Philippines was combatting an insurgency at the time of the Korean War. Immediately after the islands achieved their national independence in 1946, Louis Taruc, a popular communist leader who had been denied a seat in the Philippine Lower House, left Manila to lead his Hukbalahap guerrillas (Huks) against the new government. Taruc was aided in his efforts by a corrupt government that had only a facade of democracy to cover 'a wave of get-rich venality involving both Filipinos and Americans'.[45]

Hatred and distrust of the government grew among the disillusioned people of the Philippines. Had it not been for Taruc's own excesses of violence against the people, his Huks might have overthrown the government before 1950, when President Quirino appointed Ramon Magsaysay as his minister of defense. As defense minister and later as president, Magsaysay fashioned a combination of counter-insurgency and nation assistance activities in the 1950s that are as relevant today as they were then. First, he set his own house in order, ensuring that the Defense Department was supportive of his plans and competent to carry them out. Second, he bolstered political legitimacy with government programs that responded to public needs, such as legal assistance for the poor and limited land reform. Third, he infiltrated the Huks and, using all the instruments of political warfare, won over many guerrillas. Finally, he used limited but effective military force to ferret out the last hard-core Huks.[46]

151

Magsaysay had limited but able US assistance in the person of Air Force Lieutenant Colonel Edward G. Lansdale, whose concepts of leadership are cited as an ideal for the diplomat warrior in Chapter 5. Lansdale coined the phrase 'civic action' in the Philippines and became the prototype for the socially-conscious Colonel Edward Hillandale in *The Ugly American*.[47] Lansdale understood the interrelationship between public support and legitimacy in operations other than war. He also understood the need for strategic and operational restraint to prevent collateral damage that can undermine legitimacy.

Vietnam began as another low-level counter-insurgency effort similar to that in the Philippines. Unfortunately South Vietnam had no leader similar to Magsaysay and the US commitment there escalated from nation assistance to combat despite Lansdale's advice, creating the painful lessons learned in strategic and operational restraint discussed earlier. While the debacle in Vietnam has often been attributed to nation assistance, it was the escalation to direct combat that led the US to grief there, not the initial commitment to provide nation assistance. Strategic errors involving issues of legitimacy and the use of force determined the tragic outcome in Vietnam.

The evolving role of US Special Forces (SF) from advisors to fighters reflected the changing role of the use of US military force in Vietnam and its relationship to military and political legitimacy. Initially the focus of SF was advisory, with the emphasis on winning public support ('hearts and minds') for the South Vietnamese government (RVN) and its military forces. To their credit, SF advisors were successful with counter-insurgency and CA activities. In fact, it was their success training the Civilian Irregular Defense Group forces (CIDG) that caused the North Vietnamese to escalate the conflict, which in turn prompted President Johnson to send in the Marines in 1965. Unfortunately, the CIDG which had been effective in the guerrilla warfare against the Viet Cong was no match for the North Vietnamese regulars.[48]

Once US Marines directly engaged the North Vietnamese and the Viet Cong, it became a US war; from then on there would be no substitute for military victory. But as Philip Caputo's frustrations recounted in Chapter 1 illustrated, military victory was impossible given the political constraints imposed. Ultimate success still depended upon the public support of the South Vietnamese people, but the measure of success had shifted from winning hearts and minds to a more quantitative measure: the body count. Civilians once considered essential to mission success became obstacles to combat operations or, as Caputo discovered, even the enemy, in the dense fog of that convoluted war.[49]

Civil affairs units were utilized in Vietnam, but it was too little too late. The Marines were first with their Combined Action Platoon Program in 1965, followed by Army CA units in 1967. The effectiveness of CA combat support operations in mobilizing public support for military and political legitimacy, however, was largely neutralized by a combination of collateral damage and government corruption. When Robert W. Komer took charge of the Civil Operations and Revolutionary Development Support (CORDS) program in 1967 and for the first time managed to co-ordinate all military and civilian agencies in nation assistance, the battle for legitimacy had been irretrievably lost and with it any hope of US victory in Vietnam.[50]

Vietnam represented nation assistance gone awry. It remains a lesson in legitimacy for policymakers in the new strategic environment. The turning-point in Vietnam – the transition from nation assistance to war – came when US combat forces were committed to bolster a South Vietnamese government that lacked the military and political legitimacy to prevail. It was not the first time, nor the last, that US strategists forgot the rule that no amount of military force can substitute for a lack of legitimacy.

Cold War nation assistance was conducted during the Vietnam War under the rubric of counter-insurgency, quietly but successfully deterring communist-inspired civil war in other countries in South-east Asia, Latin America and Africa. The military instrument for Cold War operations other than war was the Special Action Force, or SAF; it was tailored around a Special Forces Group (SF Gp) and assigned the mission of promoting military and political legitimacy in its region.

The success of the SAFs was overshadowed by Vietnam, but they remain a military model relevant to the contemporary security environment. The *modus operandi* of the SAFs integrated CA into a co-ordinated nation assistance effort: SF personnel advised their indigenous counterparts on the lethal aspects of political warfare, CA and medical personnel focused on civil-military and civic action activities, psychological operations (PSYOP) forces provided the military media, and Military Intelligence personnel met intelligence requirements.[51]

Special Action Force Asia, or SAFAsia, was built around the 1st SF Gp, but had attached to it CA, PSYOP, military intelligence (MI) and medical elements. Once special operations in Vietnam shifted to the 5th SF Gp, SAFAsia devoted its attention to nation assistance in the rest of South-east Asia. In the Philippines SAFAsia worked with counterparts in the Philippine Army to continue Lansdale's policies until the early 1970s, when SAFAsia and the other SAFs were disbanded as part of the reduction in force following the Vietnam War.

In 1980 the election of President Reagan marked the beginning of a revitalized military and a new era in military strategy. Latin America became the focus of nation assistance, and CA personnel provided valuable leadership: military civic action projects were led 'by highly trained CA personnel, who can interface effectively with tactical planners, local civilian leaders and mid and high level officials of government ministries'.[52] As discussed earlier, nation assistance and CA activities remain a continuing priority for USSOUTHCOM.

Post-combat civil affairs in Grenada, Panama and Kuwait complemented continuing nation assistance activities. The military interventions in Grenada (*Urgent Fury*) in 1983, Panama (*Just Cause*) in 1989, and *Desert Shield/Storm* in 1990/91 were far more dramatic than low-profile nation assistance activities and created issues of legitimacy with the use of force. In addition to the lessons learned on unity of effort and restraint discussed earlier, these operations also confirmed the value of CA both during and following combat operations. CA helped make the transition from military to civilian control and assisted the fledgling governments in providing essential services to civilians dislocated by combat operations.

In Grenada, CA personnel arrived early and helped mitigate the effects of collateral damage. During the brief hostilities, overzealous troopers of the 82nd Airborne Division had commandeered privately-owned vehicles and modified them into armored vehicles by cutting off their tops and mounting machine guns. Claims and solatia payments were used creatively to compensate the owners of these customized cars. Following hostilities, CA projects were initiated to rebuild and improve the civilian infrastructure. In one such project, CA personnel worked with the Grenadian government and the United States Agency for International Development (USAID) to train Grenadians in construction skills in building improved school facilities.[53]

In Panama, CA personnel once again proved to be force multipliers. Here they were involved from the beginning, arriving in the airborne assault to prevent civilian interference with combat operations, and then protecting civilians in the aftermath. They remained to provide a variety of civil administration functions, such as assisting the new government in rebuilding its law enforcement and judiciary systems after ousting General Noriega's cronies.[54]

During *Desert Shield/Storm*, CA personnel became even more involved in civil administrations. The Kuwaiti Task Force (KTF) mentioned earlier was made up of senior CA officers who worked with the US State Department and the Kuwaiti government in exile to prepare for its return to power. While CA personnel did not have the long-term relationship with the Kuwaiti government as in

Grenada and Panama (Kuwait was unique in that it had the economic resources necessary to rebuild), the KTF CA advisors had a positive influence on Kuwaiti political leaders and helped smooth the transition from war to peace. Following *Desert Storm*, CA personnel in *Provide Comfort* demonstrated perseverance as discussed above, providing humanitarian and security assistance to the Kurds in Northern Iraq.[55]

Humanitarian assistance in Somalia presented CA challenges for combat soldiers. In the first phase, *Restore Hope*, there were few CA personnel involved but Army combat soldiers utilized CA techniques to achieve limited political and military objectives. In the next chapter General Edwin J. Arnold, commander of Army forces in *Restore Hope*, emphasizes the importance of CA in such humanitarian operations and recommends training Army leadership in the CA skills required of the diplomat warrior. But as described earlier, the UN nation-building phase of the Somalia intervention failed for reasons similar to Vietnam: civil war precluded effective nation assistance and political constraints precluded the effective use of combat force.

Creating order out of chaos in Haiti reflected the tension between exercising restraint and providing security through law and order. In *Uphold Democracy* CA direct support teams worked closely with Special Forces and Psychological Operations personnel to provide security to over 600 rural villages. In spite of frustrations such as those described earlier in this chapter, these teams have been successful. As one DOD official put it, they 'skillfully established the law west of the Pecos putting local thugs out of business'.[56] A ministerial support team of CA lawyers and judges addressed longer term measures '... to establish an effective judiciary in Haiti, one that will live by the rule of law rather than live in the shadow of corruption and fear'.[57]

THE FUTURE OF OPERATIONS OTHER THAN WAR: TWO VIEWS

The wrong mission

In spite of the above lessons learned on the limitations of combat operations and the value of civil-military operations during peacetime, some traditionalists still argue that the military should limit its activities to combat. Karen Elliott House has argued that operations other than war are the wrong mission, citing the depletion of funding for training and readiness, the effect of the debacle in Somalia on public support, and the supposed denigration of combat skills:

> Worst of all, nonmilitary missions eventually destroy the fighting capability of a military force. Armies, in the end, are largely composed of young men and women in uniform, not diplomats and philosophers. Such young men and women can be trained as soldiers or as policemen or as social workers, but not as all three.[58]

Ms House has correctly noted funding problems that compromise combat readiness and the importance of public support for military legitimacy, but she has failed to appreciate the strategic need for civil-military capabilities and the potential of military leaders to conduct them. Furthermore, she questioned the theological underpinnings of military professionalism by suggesting that practising the golden rule is inconsistent with warrior skills. To the contrary, it is a value embedded in Army tradition.[59]

While history has taught that civil-military operations do not degrade combat skills, Ms House has overlooked the fact that non-combat forces have (or should have) the primary role in non-combat operations. These are support and service support personnel such as civil affairs, military police, engineers, medical personnel, transportation personnel, and military lawyers, whose readiness for wartime contingencies is enhanced by their participation in operations other than war. In short, a capability for operations other than war complements and enhances the US combat capability. The painful lessons learned from Vietnam to Somalia were the result of the wrong application of warfighting, not of operations other than war.

For those like Ms House who mistakenly believe that operations other than war represent a novel and incongruous use of military forces, General Joulwan has reminded them that such non-combat operations are closer to the historic norm than combat:

> One might say that the US military is returning to normal at the conclusion of the anomalous Cold War era because, historically, normal operations for US Forces are operations other than war.[60]

The right mission

The Army Chief of Staff, General Gordon R. Sullivan, has argued that operations other than war are the *right* mission for the new strategic environment. He has described characteristics of that environment that emphasize the importance of military legitimacy: long-term political objectives that can be undermined by the inappropriate use of military force and an array of governmental and non-governmental actors that create new complexities and ambiguities for military operations.[61]

156

General Sullivan acknowledged tension between the moral and practical dimensions of US national security policy, but noted that moral interests cannot be ignored. He chided those such as Ms House who reject humanitarian assistance as a military mission:

> The expressions of surprise from some quarters that we would use military force in support of humanitarian goals ignore our history ... [T]he US Army accepts the linkage of moral and practical interests as a given ... Support of humanitarian goals is part of our past, our present, and undoubtedly our future ... New democracies are generally challenged to develop democratic institutions within their own historical contexts, to develop the role of their army in a democracy, and to define the right of minorities ... The political task at hand is to foster democratic governing structures that permit ethnically heterogeneous states to function. Our solution is federalism; we need to learn and understand what relationships will work in other countries.[62]

General Sullivan has recommended *political soldiers* to support US governmental agencies in a variety of constructive ways:

> The military obviously can provide security; it reflects our purpose for existing. But our fighting forces also can provide medical treatment; build roads, buildings, and ports; and deliver a variety of supplies, to name but a few tasks ... Both leaders and soldiers in these environments must be experts at their traditional skills but also be adept at anticipating, reading, and reacting to the complex environment ... They must understand the nuances of changing military, political, economic, and cultural dimensions and have the agility to alter our military actions quickly in a dynamic environment.[63]

PEACETIME ENGAGEMENT AND OPERATIONS OTHER THAN WAR

The real issue framed by Ms House and General Sullivan is whether the President is to have access to military capabilities beyond those of combat in conducting US foreign policy. In his National Security Strategy of Engagement and Enlargement, President Clinton emphasized the promotion of democracy and peace operations. These objectives were translated into military capabilities in the 1995 National Military Strategy, in which General John M. Shalikashvili, Chairman of the Joint Chiefs of Staff, called for a balance of non-combat and combat capabilities: 'The challenge of the new strategic era is to selectively use the vast and unique capabilities of the Armed Forces to advance national interests in peacetime while maintaining readiness to fight and win when called upon.[64] Operations other than war provide the elements of *peacetime engagement*, one of the three major components of US National Military Strategy:

> Peacetime engagement describes a broad range of non-combat activities undertaken by our Armed forces that demonstrate commitment, improve collective military capabilities, promote democratic ideals, relieve suffering, and enhance regional stability. The elements of peacetime engagement include military-to-military contacts, nation assistance, security assistance, humanitarian operations, counterdrug and counterterrorism, and peacekeeping.[65]

The strategic lessons learned from US military history, especially since Vietnam, have taught that operations other than war – by whatever name – are essential to protect US national interests in peacetime. They are the strategic elements of peacetime engagement, and most are civil-military operations; even military-to-military contacts in emerging democracies are intended to improve civil-military relations.

Much of the opposition to operations other than war has to do with their unconventional nature; during the Cold War they were considered special operations in LIC. As reflected in their doctrinal principles, they are significantly different from conventional combat operations and require unique leadership that combines the characteristics of both the warrior and diplomat. The debate over operations other than war and leadership will shape concepts of might and right in the new millennium. If critics prevail, the US will have lost a valuable capability to avoid war.

SUMMARY

Lessons learned from Vietnam to Somalia have validated the requirements and principles of military legitimacy and the need for diplomat warriors in operations other than war. These painful experiences have demonstrated the limitations of combat force in peacetime, requiring a reversal of traditional military priorities: civilians considered obstacles to combat operations are the objectives of civil-military operations, and the force required for military victory in warfighting can undermine public support for political objectives. The lethal force needed to provide security for both military and civilians alike must be balanced with restraint to ensure military legitimacy and mission success.

Ignoring the requirements of military legitimacy can convert military victory into political defeat.

Civil-military relations are an index of military legitimacy in operations other than war, and civil affairs is the military interface with civilians – a force multiplier that has proven its value in war and peace. Before the above lessons learned in legitimacy and leadership can be of any use, however, they must first be translated into military capabilities – the subject of the following chapter.

NOTES

1. Colonel Summers, Jr. has noted that 'public support must be an essential part of our strategic planning, and ... Congress has the constitutional responsibility to legitimize that support. See Harry G. Summers, Jr., *On Strategy: the Vietnam War in Context* (Carlisle Barracks, PA: Strategic Studies Institute, US Army War College, 1989), p. 4. A. J. Bacevich has also noted the importance of public support for military operations in 'New Rules: Modern War and Military Professionalism', *Parameters* (December 1990), pp. 12, 19, 22.
2. George A. Joulwan, 'Operations Other Than War: A CINC's Perspective', *Military Review* (February 1994), p. 9.
3. Ibid. at p. 5.
4. Carnes Lord, Project Director, *Civil Affairs: Perspectives and Prospects* (draft, February 1993), Institute for National Strategic Studies, National Defense University, p. 7 (hereinafter *CA Perspectives*).
5. Ibid. at p. 10.
6. Joulwan, n. 2 supra, p. 8.
7. *War Powers Resolution* (or Act), 50 USC 1541–1548.
8. FM 100–23, pp. 1–16 – 1–18.
9. Joulwan, n. 2 supra, p. 9.
10. Summers, n. 1 supra, p. 1.
11. Ibid. at p. 49.
12. Idem.
13. Summers, n. 1 supra, p. 55.
14. 'The military operation was highly successful but the civil-military operation much less so ... and the outlook for democracy and prosperity in Panama remains uncertain.' *CA Perspectives*, n. 4 supra, p. 6.
15. David L. Marcus, 'Panama: Still in Turmoil', *Dallas Morning News* (reprinted in *The State*, Columbia, SC, 16 December 1990, p. D-1).
16. In one incident a Marine gunnery sergeant shot a Somali boy with his grenade launcher after the boy reached into the sergeant's patrol vehicle and grabbed his sunglasses. For his indiscretion the sergeant was tried and convicted of aggravated assault by general court-martial. In a second incident Marine snipers firing long-range at a man with a machine gun in one of Mogadishu's busiest intersections accidentally killed a pregnant Somali woman. The shooting was in compliance with UN ROE (carrying a machine gun was evidence of hostile intent), but the incident illustrated the limitations of ROE

as a standard of legitimacy in operations other than war: ROE can never be a substitute for good judgment. Commenting on the first case, Jim Hoagland has questioned reliance on combat forces in humanitarian operations, suggesting that specially trained units be used in such ambiguous and unforgiving environments. See Jim Hoagland, *The Washington Post*, 15 April 1993, p. A-29; on ROE in Somalia, see Dwarken, *Rules of Engagement: Lessons from Restore Hope* at n. 33 to Chapter 3, supra.

17. Tobias Wolff, 'After the Crusade', *Time* (24 April 1995), p. 46.
18. See James D. Delk, 'Military Assistance in Los Angeles', *Military Review* (September 1992), p. 13. The ROE used in the LA riot are provided in the *Operational Law Handbook* (JA 422, 1993) prepared by the Center for Law and Military Operations and the International Law Division at The Judge Advocate General's School, US Army, Charlottesville, VA, at p. H-106.
19. *CA Perspectives*, n. 4 supra, pp. 7–8.
20. Bob Shacochis, 'The Immaculate Invasion', *Harper's Magazine* (February 1995), pp. 44, 54–62.
21. On civil-military relations in Latin America, see Gabriel Marcella, 'The Latin American Military, Low Intensity Conflict, and Democracy', *Winning the Peace: The Strategic Implications of Military Civic Action*, edited by John W. DePauw and George A. Luz, (Carlisle Barracks, PA: Strategic Studies Institute, US Army War College), chapter 4. On the Peru model, see Jeffrey F. Addicott and Andrew M. Warner, 'Promoting the Rule of Law and Human Rights', *Military Review* (August 1994), p. 38.
22. See Jacob W. Kipp, 'Civil-Military Relations in Central and Eastern Europe', *Military Review* (December 1992), pp. 27, 35.
23. Idem.
24. Charles E. Heller, *Twenty-First Century Force: A Federal Army and A Militia*, (Carlisle Barracks, PA: Strategic Studies Institute, US Army War College, 1993), p. 64.
25. Harry F. Walterhouse, *A Time to Build* (Columbia, SC, The R. L. Bryan Company, University of South Carolina Press, Columbia, SC, 1964), p. 56. For a shorter but more recent history of civil affairs, see Stanley Sadler, 'Seal the Victory: A History of US Army Civil Affairs', *Special Warfare* (Winter 1991), p. 38.
26. 'The parts do not exist on their own, but accept subordination to the whole. Neat lawns surround compact, trim homes, each identified by the name and rank of its occupant. The buildings stand in fixed relationship to each other, part of an overall plan, their character and station symbolizing their contributions, stone and brick for the senior officers, wood for the lower ranks. The post is suffused with the rhythm and harmony which comes when collective will supplants individual whim ... There is little room for presumption and individualism. The unity of the community incites no man to be more than he is. In order is found peace; in discipline, fulfillment; in community, security. ... West Point embodies the military ideal at its best; Highland Falls the American spirit at its most commonplace.' Samuel P. Huntington, *The Soldier and the State* (Cambridge, MA, The Belknap Press of Harvard University Press, 1957), p. 465.
27. The early Corps of Engineers made significant contributions in opening the US frontier. Among the first were Captain Meriwether Lewis and Lieutenant William Clark, who surveyed and mapped the area from the Mississippi to the Pacific Ocean. Pushing ahead of the settlers moving West, the US Army established a series of fortifications and connecting roads to provide security

against the Indian menace. Outside the stockades garrisoned by the Army early settlements were established. The Army provided the security essential to survival in the hostile environment of the 'wild west'.

Following Lewis, Clark, and Pike, Captain Benjamin Bonneville and his men, disguised as fur traders, set out in 1832 to study the habits of the Indians. The mission to the Rockies lasted five years and provided valuable data for military operations in the West. Shortly thereafter, Lieutenant John C. Fremont explored and mapped routes in Oregon territory and California. The Army also made transportation possible for early settlers, surveying and building canals, mapping railroad routes, building bridges, and even launching steamboats – all while fighting the ubiquitous Indian. In all of these undertakings the Army managed to be self-sufficient in an inhospitable land.

Back in the new capital, Captain Montgomery Meigs and his fellow Army engineers were designing and supervising the construction of public facilities, including the Washington aqueduct. Over the years the Army Corps of Engineers was responsible for building the Washington Monument, buildings to house the State, War, and Navy Departments, the Library of Congress, the Old National Museum, the Old Post Office, the Pentagon, and many bridges, schools, streets, and other public facilities. Walterhouse, n. 25 supra at pp. 57–58.

28. Ibid. at pp. 58, 78 (note 42).
29. Ibid. at p. 59.
30. Idem.
31. Ibid. at p. 60.
32. Ibid. at p. 62.
33. A common saying of the Depression era indicated the need for erosion control on land wasted by overplanting cotton: 'General Erosion has wrought more havoc on the land than General Sherman'. One of the few surviving CCC commanders, Colonel George Buell, 91, of Summerville, South Carolina, remembers the hectic days of 1933 when the Army was given three months to prepare for the onslaught of unemployed boys and men: 'It was astounding what was done overnight. The Army had these warehouses that were full of everything that was needed – clothing, cots, tools, tents . . . everything.' Glenn Oeland, 'Heritage of Hard Times', *South Carolina Wildlife*, July–August 1992, pp. 16, 23.
34. One CCC recruit wrote home of the experience: 'During the first few days the company did little except eat, sleep, and take long hikes'. But then came a fight: 'The chief weapons were bottles, but any object available was put into action. No deaths or bodily injury occurred, but the next day the company was sent to the woods to work, and that work hasn't stopped yet.' Idem.
35. Charles E. Heller, n. 24 supra at pp. 68–69.
36. Colonel Joseph K. Dietrich, the senior USAR advisor to the United States Special Operations Command at MacDill Air Force Base in Florida has proposed that reserve component Special Forces (SF), Psychological Operations (PSYOP), and CA forces use their annual training active duty to provide military civic action in a remote Indian reservation in Montana. The citizen-soldiers would 'conduct training in teaching people to help themselves focusing on the young people with the objective of increasing pride and self-esteem.' The project is based on an earlier successful model known as Special Proficiency at Rugged Terrain Training and Nation Building (SPARTAN) which helped local residents build improvements on a reser-

vation plagued with endemic problems of alcoholism, drug abuse, a high school drop-out rate, and unemployment. The initial reaction of senior commanders to Col Dietrich's proposal was '... great but there are statutes against soldiers being involved on the domestic scene'. This is a typical reaction to innovative civil-military projects in an over-regulated environment; there are legal restrictions, but they do not preclude such projects.

37. See Ralph R. Young, *Snapshots of Civil Affairs: A Historical Perspective and Views*, unpublished paper presented at the 39th Annual Conference of the Civil Affairs Association at San Antonio, Texas, June 1986, p. 4. Also Alexander M. Walczak, *Conflict Termination–Transitioning From Warrior to Constable: A Primer*, unpublished paper prepared as part of USAWC Military Studies Program, US Army War College, Carlisle Barracks, Pennsylvania, 1992. Walczak emphasizes the responsibility of commanders for the welfare of civilians in their areas, and their role as constables in establishing and maintaining law and order and providing essential services until CA forces arrive.

38. Ted B. Borek, 'Legal Services During War', 120 *Military Law Review*, 1988, pp. 35–40.

39. US Army Field Manual (FM) 41–10, *Civil Affairs Operations*, December 1985, p. 1–1. Joint doctrine has since confirmed the priority of operational law to CA operations. See JCS PUB 3–57, *Joint Civil Affairs Operations* (Final Draft, November 1990), pp. *xx*, I–1, I–7, II–3, IV–6, and C–2, 3. For a discussion of some of the legal issues peculiar to CA and their relationship to legitimacy, see Barnes, 'Legitimacy and the Lawyer in LIC: Civil Affairs Legal Support', *The Army Lawyer* (October 1988), p. 5.

40. General William Richardson, Commanding General, Army Training and Doctrine Command, draft memorandum to General John Wickham, Army Chief of Staff, Subject: 'Civil Affairs Modernization', dated June 1986, p. 4.

41. Walterhouse, n. 25 supra, pp. 73–74.

42. Ibid. at pp. 76.

43. Ibid. at pp. 76, 77.

44. Ibid. at p. 77.

45. Ibid. at p. 84.

46. Ibid. at pp. 84–90.

47. Reference is made to William Lederer and Eugene Burdick's *The Ugly American* (New York: W. W. North & Co., 1958).

48. Charles M. Simpson, III, *Inside the Green Berets: The First Thirty Years: A History of the U.S. Army Special Forces* (Novato, CA: Presidio Press, 1983), Chapters 10 and 13; also, Shelby L. Stanton, *The Rise and Fall of an American Army* (New York: Dell Publishing Company, Inc., 1988), Chapters 1 and 2.

49. See references to Philip Caputo's experiences as a Marine Corps lieutenant in Vietnam from his book, *A Rumor of War* in Chapter 1, nn. 35–41.

50. William R. Berkman, 'Civil Affairs in Vietnam', unpublished paper written for US Army War College, Carlisle Barracks, PA, December 1973.

51. See Charles M. Simpson, III, *Inside the Green Berets*, cited at n. 48 supra, pp. 69–70.

52. John T. Fischel and Edmund S. Cowan, 'Civil-Military Operations and the War for Moral Legitimacy in Latin America', *Military Review* (January 1988), pp. 40, 43. The authors use the term civil-military operations, which has essentially the same meaning as civil affairs in this context. For other

examples of military civic action in Latin America and elsewhere, see *Winning the Peace: The Strategic Implications of Military Civic Action* cited at n. 21 supra.

53. See Delbert L. Spurlock, 'Grenada Provides Classic Case', *The Officer* (August 1984), p. 17; also, Barnes, 'Grenada Revisited: Civil Affairs Operates in Paradise', *The Officer* (July 1985), p. 14.

54. See 'Civil Affairs in Just Cause', *Special Warfare* (Winter 1991), p. 28.

55. See John R. Randt, 'Working in a Place Called Zacho: Stories from the Storm', *Army Reserve Magazine* (third issue of 1991), p. 10.

56. *Civil Affairs Journal and Newsletter* (Civil Affairs Association, Kensington, MD), January/February 1995, p. 3.

57. Bill Maddox, 'Haiti Recovers', *Army Reserve Magazine* (Spring 1995), p. 20.

58. Karen Elliott House, 'The Wrong Mission (Beyond the Cold War: Foreign Policy in the 21st Century)', *The Wall Street Journal*, 8 September 1994.

59. 'Compassion, or respect for the dignity of each individual, reflects a moral code as ancient as the *Golden Rule*, the mandate to treat others as one desires to be treated. Americans have shown their compassionate nature in peace and war ... during humanitarian missions like Provide Comfort in northern Iraq and Provide Hope in Somalia'. *Army Focus 1994, Force XXI*, An Official Department of the Army Publication (1994), p. 35.

60. Gordon R. Sullivan, 'The Challenge of Peace', *Parameters* (Autumn 1994), pp. 4, 7.

61. Ibid. at p. 8.

62. Ibid. at pp. 9–10.

63. Joulwan, supra n. 2, p. 10.

64. *United States National Military Strategy, 1995*, Department of Defense, Washington, DC, p. 1.

65. Ibid. at p. *ii*. The elements of peacetime engagement are further elaborated at pp. 8–9.

163

7

New Capabilities for the New Total Force

They will hammer their swords into plowshares,
their spears into sickles.
Nation will not lift sword against nation,
there will be no more training for war.

Isaiah 2:4

Nation will rise against nation,
and kingdom against kingdom.
There will be famines and earthquakes in various
places.

Matthew 24:7

The new and uncertain strategic environment requires flexible defense strategies and capabilities for both war and peace. It is a time for both swords and plowshares. Military operations other than war require non-combat capabilities that complement warfighting; both civil-military and military-to-military capabilities must be as constructive as warfighting is destructive. Civil-military activities require the civilian soldiers of civil affairs (CA); and military-to-military missions that promote democracy, human rights, and the rule of law offer military lawyers an expanded operational role. All require diplomat warriors who are as comfortable wielding plowshares as swords.

These concepts are recognized in military doctrine but are not yet real capabilities. If operations other than war are to be a strategic option, there must be personnel within the Total Force to conduct them according to the requirements and principles of military legitimacy. Creating these new capabilities in a time of downsizing will require changes in military priorities, and change does not come easily to the world's largest bureaucracy.

One strategist aware of this bureaucratic inertia has predicted that

Congress will rely on the reserve components for domestic and civil-military activities:

> Given the current domestic situation, it is becoming increasingly clear to members of Congress, Senator Nunn for example, that there is a significant role for the military to play in accepting peacetime missions that can lead to strengthening America. There is a strong institutional bias within the Army against involvement in such missions.
>
> ... in the future, it is the USAR portion of the Federal Army that is structured and deployed to take on the domestic nation assistance directed by the Congress and the President.[1]

CIVILIAN SOLDIERS AND CIVIL AFFAIRS

If Congress expands the peacetime role of the civilian soldiers of the United States Army Reserve (USAR) and Army National Guard (ARNG) and effectively integrates them into a seamless Total Force, it will not only provide the resources required for operations other than war, but also help change the outdated paradigm of the soldier and the state.

The reservist, or civilian soldier, has always been a major component of the Total Force, but budget constraints and a new strategic environment that emphasizes civil-military relations require even more effective integration and utilization of civilian soldiers in the new Total Force.

Civilian soldiers are the bridge between the authoritarian military and the civilian society it must serve. They are the lubricant that eases the friction of civil-military relations. They also provide many of the civilian skills needed by diplomat warriors in operations other than war.

Reservists help mobilize public support required for the legitimacy of peacetime military activities. Any time reservists are part of military operations, so is the American public; and assuming the other requirements of military legitimacy are met, public support is assured. This was evident in the strong public support for *Desert Shield/Storm* which relied heavily on the reservists, in contrast to the lack of public support for the Vietnam conflict where relatively few reservists were involved. *Restore Hope* was more like Vietnam than *Desert Shield/Storm*.

Thousands of reservists were mobilized for the Gulf War, but such mobilizations are not practical for operations other than war. Reservists must be more effectively integrated in the new Total Force, not only as a contingent wartime combat reserve but also as part-time peacetime diplomat warriors. In fact, the term 'reservist' is a misnomer for those civilian soldiers with continuing peacetime missions. They are actually part-time soldiers in the front lines of operations other than war, in contrast to their counterparts in the combat arms who are true reservists for wartime contingencies.

Any US capability for operations other than war must rely on CA civilian soldiers for civil-military activities, since 97 per cent of the CA force structure is in the Army Reserve.[2] CA diplomat warriors bring to peacetime operations valuable skills in law enforcement (public safety), public health, public administration, public relations, engineering, civil law and religion. These are only seven of the 20 CA functional areas that mirror essential government services that are critical in nation assistance.[3]

The combination of political and diplomatic skills with traditional military leadership make CA officers well suited for all civil-military activities. Within the Army officer corps they exemplify the diplomat warrior and give credence to the new paradigm of the political soldier. In transitional nations they can help improve civil-military relations by advising their counterpart officers on the role of the military in a democractic society.

CIVIL AFFAIRS MISSIONS IN OPERATIONS OTHER THAN WAR

In wartime and conflict, CA supports combat forces; but in peacetime priorities are most often reversed: combat forces should support CA in civil-military operations. Whenever civil-military relations and legitimacy are essential to mission success, CA is more than a force multiplier: it is the military's link with the public support required for legitimacy. This is evident in three categories of CA activities: combat support and post-conflict activities, continuing nation assistance and emergency nation assistance.

Combat support and post-conflict activities

CA combat support reflects the wartime primacy of force over political objectives. Its focus is on preventing civilian interference with combat operations and providing civilian support (labor and supplies) for military operations, while ensuring command compliance with legal and moral obligations to civilians. Post-conflict CA activities reflect the priorities common to operations other than war, with political objectives predominating over the use of force. They usually begin as emergency humanitarian assistance and evolve into continuing nation assistance activities.[4]

Provide Comfort was a post-conflict operation following *Desert Storm*; it began by helping the Kurds in northern Iraq through disaster relief and refugee control, and evolved into longer term humanitarian and security assistance (it would be called nation assist-ance if Kurdistan existed as a nation). *Provide Comfort* illustrated the value of CA expertise in achieving military legitimacy:

> In this case, the military commanders conceived and planned the operation as a fundamentally civil-humanitarian operation carried out by both military forces and civilian agencies (both US and inter-national). While the need to ensure security for the Kurds was a major consideration, military issues were never at the forefront. The primary focus was on humanitarian assistance activities to feed, house, and care for Kurds displaced from their homes by Saddam's campaigns.... One major reason this operation was carried out more smoothly than its larger counterpart in Kuwait and southern Iraq was the availability to the European-based commanders of expert CA advice from trusted members of the team.[5]

General Wayne A. Downing, Commander of the United States Special Operations Command (USSOCOM), has emphasized the importance of the CA post-conflict capability to secure the victory:

> Any operation we undertake in the future will have to include civil affairs. While we have always recognized the moral and legal obliga-tions of the commander to the civilian population, the impact of this role has grown in recent years. Such challenges as dealing with refugees and cementing military victory with a plan to create stable nations in the aftermath of war highlight the importance of civil affairs to the commander. We must not only win the war, we must win the peace. Civil affairs is a key part of this post-conflict mission.[6]

Continuing nation assistance

As described in Chapter 2, nation assistance includes a wide range of humanitarian and security assistance missions conducted under the auspices of the US ambassador. As noted earlier, General Joulwan has practised unity of effort by working closely with US ambassadors in USSOUTHCOM. He has recognized CA as the core of the USSOUTHCOM nation assistance capability:

> ... provding assistance to developing democratic governments by co-ordinating and conducting civil affairs training assistance visits; providing humanitarian and disaster relief assistance when required; and supporting the development of viable host nation counter-insurgency programs.[7]

Emergency assistance

Spreading primal violence has created floods of refugees across Africa, Eastern Europe, and closer to home in Haiti. Military operations such as *Provide Comfort, Restore Hope*, and *GITMO* have attempted to contain this human flood tide, and their legitimacy has depended upon the protection of human rights and the humanitarian treatment of refugees.

Refugee control and disaster relief are emergency nation assistance activities that have recently proved to have a high priority, whether or not associated with combat. Jim Hoagland of the *Washington Post* noted their importance, and made the unlikely comparison of refugee control to nuclear arms control:

> Combatting population flight is likely to occupy global strategy in this decade much as nuclear arms control and international arms integration did in the 1980s. It should be an organizing principle of international action beyond the Cold War.... Foreign aid, trade concessions, investment strategies, and even military intervention are becoming tools to achieve this goal.[8]

Disaster relief is a surge form of humanitarian assistance. Mr Hoagland noted that in May 1991, 20,000 US forces were deployed to help disaster victims in Bangladesh, as well as Kurdish war refugees in *Provide Comfort*. He called for a new military capability to provide disaster relief, but acknowledged bureaucratic resistance:

> The military forces of the world's great powers organize in peacetime for every contingency however unlikely. They are the ideal core of a new disaster relief system. ... [But] Generals and international-aid bureaucrats will balk at the concept of institutionalizing the military role in relief. This is not what military units are for, they will argue. But in the post-Cold War era there could be no better rationale for keeping a significantly sized military force with global logistical capabilities.[9]

Civil affairs has a pre-eminent role in emergency assistance. One authority has described CA as

> the only part of the military force structure prepared by doctrine, training, experience, and personnel recruitment policy to deal with these organizations [that take the lead in humanitarian relief, and whose managers] ... have repeatedly commented how well they could work with US forces if they could deal with civil affairs officers instead of combat commanders.[10]

Changes needed in the Total Force

A CA capability for combat support and limited post-conflict activities already exists; but, as suggested by Jim Hoagland, providing a military capability for emergency and continuing nation assistance activities will require changes that will not come easily. These changes should begin at home, where there is a need to realign the missions and capabilities of the ARNG and USAR.

Domestic assistance: realigning missions and capabilities

The ARNG is the primary source of emergency military assistance to domestic civil authorities, and in recent years has proven indispensable to governors facing natural and man-made disasters. The military was so effective with disaster relief following Hurricane Hugo in South Carolina in 1990 and Hurricane Andrew in Florida in 1992 that some policymakers have argued that the primary responsibility for managing domestic disaster relief should be transferred from the Federal Emergency Management Agency FEMA to the Department of Defense (DOD).[11]

Colonel Charles Heller has suggested that the American public would be better served if the United States Army Reserve Command (USARC) in Atlanta assumed the functions of the FEMA bureaucracy:

> The taxpayers would certainly be pleased that not only could the Federal Government be reduced by one bureaucracy, but also happy that defense dollars spent on the USAR would have double value by providing them with the support needed in times of national domestic emergencies.[12]

Even if the USARC assumed the FEMA function of co-ordinating federal resources employed during national emergencies, the individual states would still have primary responsibility for domestic emergencies. The primary state mission of the ARNG is providing military assistance to domestic authorities in civil emergencies; and unless and until federalized, the ARNG functions under the command of the state governor.

Each state has a state area command, or STARC, to provide command and control during domestic emergencies. Federal forces often assist the ARNG through the STARC, but never come under its command. Since many of the military specialties needed during a domestic emergency are in the USAR, mobilizing the required federal forces and co-ordinating their separate chain of command with a STARC is at best unwieldy, and would continue to be so under current law even if the USARC were to replace FEMA.[13]

The specialized military capabilities needed by governors during domestic emergencies are similar to those needed in nation assistance, and share an emphasis on civil-military relations and military legitimacy. CA includes many of these capabilities in its 20 CA functional areas; they represent a full range of public services provided by state government, of which the STARC is a part.[14]

There are no CA units currently assigned to the ARNG; all are in the USAR where federal law and regulations put them effectively beyond the reach of governors.[15] Each state ARNG should have a dedicated CA unit with its functional specialties aligned with its STARC to ensure both competence in the CA function areas and good civil-military relations during domestic emergencies.

The ARNG has few of the non-combat capabilities needed for its primary state mission of disaster relief and riot control since it is primarily a reserve combat force.[16] Governors do not need tanks and howitzers for riot control and disaster relief; they need military personnel with the specialized skills required to assist civil authorities provide law and order and essential human services. These support and service support capabilities are found in the USAR, and include CA, military police, engineers, transportation, hospitals and legal units.[17]

The law also favors the ARNG for riot control. Federal forces (active or reserve) are prohibited from providing law enforcement by the Posse Comitatus Act.[18] The ARNG is not subject to these restrictions while in a state status. The contrast was evident during the 1992 Los Angeles riots: during the early stages when the California National Guard was in a state status it was able to accept 100 per cent of the mission requests received from law enforcement authorities; after it was federalized and joined by additional federal forces they could accept only 10 per cent of the mission requests.

A realignment of missions and capabilities between the USAR and ARNG is overdue, but old traditions die hard. The ARNG prefers being a combat force, but its primary mission is non-combat. Reserve combat forces fit better in the USAR since it is a federal force where training with active component units would not be complicated by governors in the chain of command. Tradition, politics and bureaucratic intransigence should not prevent realignment of mismatched missions and capabilities. These obstacles must give way to new security requirements and budgetary constraints.[19]

Nation assistance: diplomat warriors in an inter-agency command

As described in Chapter 2, nation assistance includes a wide range of continuing humanitarian and security assistance activities. These require diplomat warriors with a regional cultural and political orientation and linguistic ability. Their mission requires continuous interface with indigenous civilians and their military counterparts; as part of the US forward presence they must avoid military enclaves that have produced 'ugly Americans'; their presence must emphasize military legitimacy.

CA reservists can provide much of the capability for nation assistance (and forward presence) on a rotating basis.[20] But civilian soldiers are no substitute for career diplomat warriors who must be the full-time component of a nation assistance capability; they provide the profesionalism and continuity required to achieve long-term US political objectives. Unfortunately, while the need for these career diplomat warriors is increasing, their numbers have decreased dramatically during recent reductions in force.

Foreign area officers are the most experienced diplomat warriors in the active component, but they are not well represented in the higher echelons of the Department of Defense. As a result they have suffered disproportionate cuts before selective early retirement boards. These officers will be difficult and expensive to replace; they have developed fluency in language as well as political and cultural orientation to their specialty areas, all at a cost far greater than that required for combat officers. Unless priorities are adjusted quickly a valuable capability for nation assistance will be irretrievably lost.

In addition to the lack of diplomat warriors in the active component there is no inter-agency force structure for nation assistance. A good starting point for such a command would be the United States Army Civil Affairs and Psychological Operations Command (USACAPOC) at Fort Bragg, NC, a subordinate command of USSOCOM. USACAPOC is unique in that it has integrated active and reserve component Army personnel required for civil-military operations. Most of its personnel are in the reserve component: 97 per cent of CA personnel and 87 per cent of PSYOP personnel are reservists.[21]

Congress mandated the creation of USSOCOM in an amendment to the Defense Reorganization Act of 1986 which required a new unified command to provide a capability for non-traditional (special) military operations. Ten special operations activities were identified in the Act: they include CA, PSYOP and humanitarian assistance; all emphasize military legitimacy to achieve political objectives.[22]

The uniqueness of USACAPOC should be its strength, but has instead been its weakness. In the competition for limited resources and missions during downsizing, USACAPOC has been degraded relative to the other Army Special Operations Forces (Special Forces, Special Forces Aviation and Rangers) which are oriented to combat and have a higher percentage of active component personnel. In the active Army force structure, combat units have a higher priority than non-combat units. While this should be the rule in warfighting, it does not apply to operations other than war.

The preference for combat within DOD is understandable, but it represents an organizational bias that is a serious obstacle to an effective capability for operations other than war. Such a military capability must be closely aligned with the Department of State (DOS) since it has the dominant role in security and humanitarian assistance. But the DOS has little presence or influence in USSOCOM and is absent in USACAPOC. The principles of military legitimacy, political objectives, unity of effort and perseverance require that DOS involvement and oversight be built into force structures tailored for nation assistance.

The commander in chief of USSOCOM, General Downing, cited the importance of inter-agency co-operation in post-conflict CA activities following both *Just Cause and Desert Storm*:

> One of the great lessons of Panama was the need for inter-agency work in dealing with the manifold problems involved in returning to normalcy. One of the great success stories of KTF [Kuwait Task Force] was its work with the Office of Foreign Disaster Assistance.[23]

To overcome the institutionalized resistance to inter-agency co-ordination, a CA study suggested the creation of an inter-agency committee for civil-military operations under the auspices of the National Security Council, together with 'an Executive Order covering peacetime civil-military operations and inter-agency operations at the national and regional levels.... The study noted an absence of guidelines on inter-agency activities:

> There is no document that establishes overall guidelines for how civilian agencies and DOD should work together to conduct civil-military operations.[24]

There are endless variations of structural arrangements to reflect a joint venture between DOD and DOS in nation assistance. Whatever the structure, there must be operational support and oversight from DOS at all levels, similar to that on country teams and joint US military assistance advisory groups (JUSMAAGs). Political advisors assigned at operational levels within a new inter-agency command could provide such support and maintain continuing liaison with DOS.

Inter-agency training is also important for unity of effort. To ensure both proficiency and compatibility in inter-agency activities, diplomat warriors should be trained alongside their foreign service counterparts at the Foreign Service Institute and other DOS facilities.[25]

A forward presence for nation assistance: the security assistance force

In strategically important regions where both a forward presence and nation assistance are required, the old Special Action force of the 1960s should be resurrected to provide inter-agency command and control. The concept has been retained in current doctrine by a different name, albeit one sharing the same acronym: the Security Assistance Force, or SAF. The prime virtue of the SAF is that it has no fixed shape; it can be tailored to the unique needs of any region.[26]

CHANGES NEEDED IN THE LAW

The effective utilization of reservists as diplomat warriors in nation assistance operations will require a change in the law which currently limits reservists to training while on active duty, unless they are mobilized.[27]

General Downing noted the problem in Panama, where reserve CA units were not activated, and the improvement in *Desert Storm*:

> Our civil affairs units were simply magnificent. Their skills, talents and experience cannot be replicated in the Active Component. Unfortunately, the problem of access still remains. We still need a presidential call-up to access these talented Reserve units in time of crisis. This limits our effectiveness in dealing with the ongoing demands in peacetime and in war for the unique skills that exist only in the reserve Component civil affairs units.[28]

The laws restricting reservists to training during peacetime are an anachronism in the new strategic environment. They were intended for reserve combat forces which have a wartime contingency mission and need not become operational without a mobilization. But these laws are clearly unsuited for those civilian soldiers who must be front-line forces in operations other than war.

MILITARY LAWYERS AND MILITARY LEGITIMACY

Military lawyers in both the active and reserve components represent a largely untapped source of military advisors for nation assistance. They are an integral part of a CA capability, but they also represent an independent capability well suited to advise both military counter-parts and civilian authorities in transitional nations.[29]

By professional training and discipline, lawyers are advocates and advisors, not commanders, and are well versed in the practical application of the rule of law as it relates to human rights and democracy. Operational law (OPLAW) is the stock-in-trade of the military lawyer, and complicance with OPLAW is the largest part of military legitimacy.

Staff support for law and legitimacy

The command or staff judge advocate (SJA) is the commander's legal advisor, and is responsible for advising the commander on all OPLAW matters, including those involving civilians. On a general staff the commander also has a civil-military advisor, the civil-military officer (CMO) or G-5, who is '. . . the principal staff assistant to the commander in all matters concerning political, economic and social aspects of military operations'.[30] The CMO must co-ordinate closely with the SJA in assisting command compliance with OPLAW requirements affecting civilians.[31]

Commanders may not have access to a CMO with CA expertise; with 97 per cent of CA officers in the USAR, there are few CA officers available to serve on active component staffs. When no CA officer is available the next best CMO is a military lawyer; that is because compliance with OPLAW (as it relates to civilians) is a major part of the CA mission.

The Marine Corps provided a useful precedent for all Services when it assigned CA staff support functions to its lawyers. The CA tasks assigned to Marine Corps lawyers are not limited to legal support, but include all CA command support functions until CA personnel can assume them. Army lawyers would benefit from such broadened responsibilities; they would be command advisors on both the law and legitimacy.[32]

There are precedents within the Army for having military lawyers serve as G-5. During the Second World War approximately 200 highly-qualified lawyers were assigned to civil affairs and military government duty. Moreover, there were instances in which commanders had their SJA perform the function of the CMO, recognizing the similarity of the duties of the two staff functions. Based on this experience, recommendations were made after the war to assign CA legal duties (then referred to as military government) to the SJA.[33]

Promoting military legitimacy

While the primary function of the military lawyer is advising commanders and staff on the requirements of OPLAW, and by extension, military legitimacy, military lawyers have also demonstrated a capability to advise counterpart military lawyers and civilian authorities in emerging democracies on matters of law and legitimacy.

In Latin America, Army lawyers have led a unique joint venture with their Peruvian counterparts to develop and institutionalize training within the military for human rights. To achieve the long-term objective of helping the Peruvian military assume a more 'professional role appropriate to a democracy' the concept of human rights had to be inculcated 'into the psyche of the military'. The training was designed to 'foster greater respect for, and an understanding of, the principle of civilian control of the military; and improve military justice systems and procedures to comport with internationally recognized standards of human rights'.[34]

Over the years Peruvian soldiers had become hardened by the brutality of the *Sendero Luminoso* (Shining Path) and the drug traffickers who were their partners in crime. President Fujimori understood the threat of military brutality to military legitimacy and the need for the Peruvian military to respect human rights.

> Foremost in Peru's fight for survival was maintaining the legitimacy of the Peruvian government, wherein true democracy would have a chance to endure. A major step in remedying the legitimacy issue was to inculcate human rights and the law of armed conflict training into its armed forces.[35]

Such advisory activities to improve human rights and civil-military relations are a priority in Latin America according to General Joulwan, who has advocated ambassador-developed country plans to

> .. assist nations within the theater with the professional development of their military forces to guarantee human rights while defending against internal and external security threats.[36]

General Joulwan left USSOUTHCOM to become commander of US forces in Europe; now he will be able to apply his successful experiences in Latin America to the challenges of his new AOR. That should include adapting the Peru model for promoting human rights and civil-military relations to the emerging democracies in Eastern Europe, where, as in Latin America, the militaries have a history of human rights violations that have contaminated civil-military relations.

Army lawyers are already promoting military and political legitimacy with their counterparts in Eastern Europe. They have also helped civil authorities develop the legal structures required for democracy in Eastern Europe. In a 1992 conference hosted by the European Community for emerging democracies, Army lawyers presented classes and hosted panel discussions. The theme of the conference was *The Role of the Military in a Democratic Society* and there were particpants from Lithuania, Latvia, Estonia, Czechoslovakia, Hungary, Bulgaria, Romania and Albania. According to an Albanian delegate, military lawyers at the conference were something of a curiosity:

> Prior to the revolution we had no military lawyers in the army. We do not know what a legal advisor is because we have not had any in the past.

A Lithuanian delegate expressed reservations about civilian control of the military:

> You speak of civilian control – well, we must wait and see which direction the political leaders will take us.[37]

One of the US Army lawyers in attendance characterized the conference as a success and was optimistic about promoting democratic ideals in the region:

> An idea implanted now may find its way quite easily into a constitution, a presidential directive, or a national regulation and may reap enormous benefits in the years to come.[38]

CA legal support

Army lawyers have an especially important role in CA due to its emphasis on compliance with OPLAW; but more is expected from the CA lawyer than legal support. Military legitimacy requires an understanding of how the law relates to political objectives and civil-military relations. The CA lawyer must be proficient in more than OPLAW; he or she must understand how the rule of law gives meaning to democracy and human rights.

The pivotal role of lawyers in matters of legitimacy is reflected in a CA study that recommends *Expanded Political and Legal Missions for CA*:

> The experience of recent conflicts has validated an expanded role for CA in certain sensitive political and legal areas. These CA missions need to be given a clearer foundation in doctrine and incorporated in joint planning and training.

The study noted that human rights were a major concern in newly-liberated areas during the Gulf War, and that

> ... the broad area of moral and legal responsibilities toward the civilian population in a conflict environment calls for a significant role for CA personnel.

It underscored the need for CA legal support operational functions (for example, helping establish a court system) to co-ordinate closely with command legal support staff functions:

> Ideally, CA and JAG [Army lawyer] functions should mesh closely and complement one another, with CA units acting as an operational arm of senior command echelons.

For political and legal missions such as those involving the promotion of democracy, human rights and the rule of law, the study recommends

> ... the direct interaction of CA officers with US ambassadors as well as most senior officials of foreign governments. The CA role in legal aspects of humanitarian issues needs careful examination, with particular emphasis on the relationship of CA and JAG roles and missions. Such a review ought to consider current requirements for and possible enhancements of legal expertise within the CA force structure.[39]

Military legitimacy and military justice

Military lawyers have traditionally focused their time and efforts on military justice matters (i.e., military criminal prosecution), not on OPLAW support. But these priorities are likely to be reversed by the increased OPLAW requirements for operations other than war, coupled with reduced requirement for courts-martial in the all-volunteer peacetime military.

As suggested in Chapter 5, civilian soldier lawyers and judges could assume a greater role in the trial of courts-martial, freeing active component lawyers to focus on operational issues. Other benefits would include reducing negative public perceptions of the court-martial as a military court, minimizing command influence and providing significant defense savings.[40]

With increasing pressure to reduce defense costs by eliminating non-essential services, the peacetime court-martial should become the primary responsibility of those reserve component lawyers and judges who must staff it in wartime. These reservists could try the relatively few peacetime courts-martial on a part-time basis, and in so doing ensure their readiness for wartime responsibilities and eliminate the need for an expensive full-time military infrastructure dedicated to criminal prosecutions.

Military legitimacy would also benefit from such a restructuring of the military justice system. Civilian soldier lawyers and judges who regularly practise in civilian criminal courts would bring to the military courtroom a healthy mix of civilian and military values, helping moderate negative public perceptions of a separate but equal system of military justice.

LEADERSHIP TRAINING IN MILITARY LEGITIMACY

The diplomat warriors of CA and military lawyers are not the only officers who must understand the concept of military legitimacy. Leadership training in the requirements and principles of military legitimacy must be provided to all military leaders involved in civil-military operations. Major General Edwin J. Arnold, Jr., commander of the 10th Mountain Division (LI) during *Restore Hope*, has advocated training for senior leaders and staff which emphasizes ROE and cultural issues which

> ... focus on such requirements as negotiations, UN operations, integration of all services and coalition forces, inter-agency operations, and operating with non-governmental organizations.[41]

Training in military legitimacy should be an element of leadership training from ROTC through senior service schools. The diplomat warrior model of leadership and principles of operations other than war should be taught along with the principles of warfighting, and legitimacy emphasized as an integral part of instruction in military law and justice. Unfortunately, the concept of legitimacy is not included in ROTC training and the quality of instruction in military law is not what it should be.

West Point provides the ideal venue for legal instruction. Each cadet receives over 50 hours of legal instruction provided by a faculty of 15 of the brightest military lawyers in the Army. By way of contrast only 17 hours of instruction in military law and justice are required in ROTC during the entire four-year curriculum, and there are no military lawyers assigned to any ROTC detachment. Army ROTC is the largest of all pre-commissioning programs, but an Army regulation that requires that military lawyers teach military justice to officer candidates has never been applied to ROTC.[42]

A prototype ROTC Legal Instruction Project co-sponsored by the Department of Law at West Point, Cadet Command, and the Army Judge Advocate General's School has introduced the concept of military legitimacy as part of an expanded curriculum on military law and justice.[43] The project represents both a foundation for Army officer training on military legitimacy, and an opportunity to reduce the disparity between the quality of legal instruction provided to ROTC cadets and cadets at West Point.

Funding constraints have stalled the project. Its implementation would require USAR military lawyers teaching at over 300 ROTC detachments across the country. There is no shortage of reserve component JA officers to teach ROTC, but for reasons unknown, no reservists other than Active Guard and Reserve (AGR) officers on extended active duty have been assigned to teach at ROTC detachments.

In a time when reservists should be assuming as many active component functions as possible, they should be teaching ROTC cadets; after all, the 'R' in ROTC stands for *Reserve*. This should not be limited to lawyers – there are many qualified instructors in the USAR and ARNG that could teach ROTC courses.[44] But it is especially important that military lawyers teach military law to officer candidates – not only to comply with Army regulations, but also to ensure that future Army leadership understands both the law and legitimacy.[45]

At more senior levels, leadership training should address more complex issues of legitimacy in operations other than war. The Commandant of the Army War College has emphasized the need for training in negotiation skills and rules of engagement, topics well suited for military lawyers.[46] The inter-agency dimension of operations other than war also requires that civil-military relations be better understood as a command responsibility, along with CA roles and missions. Currently there is little understanding of CA within DOD, the inter-agency arena, and in the field.[47]

General Arnold's experience in Somalia convinced him of the relevance of civil-military relations to mission planning and execution. His command worked among and through Somali social structures, which included 21 clans and sub-clans. Civil affairs teams and other soldiers used diplomacy and negotiating skills to deal with the many different, and sometimes hostile, clans and elders in each village.[48]

Civil-military operations centers were established to achieve unity of effort between military and non-governmental organizations in different sectors of Somalia. Inter-agency relations with the State Department and the UN command authority (UNOSOM) required a close working relationship with the Ambassador, but there were inadequate State Department personnel in country to do the embassy's work. The shortage of CA personnel required combat leaders to fill diplomatic positions in the humanitarian relief sector.[49]

General Arnold concluded that it is not practical to rely on limited

CA forces in operations other than war. He has advocated special training for combat leaders to prepare them for CA missions, and elaborated on the topics that should be included in staff and leadership training:

> We should consider adding to that training topics such as coalition warfare, negotiations, civil disarmament, extensive urban operations, operating with non-governmental organizations, interagency operations, co-ordinating with State Department and UN personnel, and dealing with the complexity brought about by operations other than war.[50]

Since it may be unrealistic for commanders to expect CA units or even CA staff support whenever needed, it is especially important that senior officers understand the requirements and principles of military legitimacy as they relate to civil-military relations. In short, they must become diplomat warriors.

SUMMARY

The new Total Force must include non-combat capabilities for operations other than war to complement its warfighting capabilities. These capabilities must rely on more effective integration and utilization of civilian soldiers and military lawyers in a seamless Total Force. The following recommendations incorporate the requirements and principles of military legitimacy in these capabilities:

• An interagency CA command should be created to provide the specialized units and personnel required for domestic emergencies and nation assistance.

• The combat and non-combat forces of the USAR and ARNG should be realigned to match capabilities with missions.

• The law should be changed to allow reservists with the specialized skills needed in operations other than war to become operational as part-time diplomat warriors without first having to be mobilized.

• Military lawyers should be utilized as civil-military officers to advise commanders on matters of military legitimacy, to advise both military counterparts and civilian authorities in emerging democracies on democracy, human rights, and the rule of law, and to assume military justice functions (the court martial) to allow their active component counterparts to focus on OPLAW issues.

• Military legitimacy should be incorporated in leadership training from ROTC to senior service schools, with reservists, especially military lawyers, assuming a greater role in teaching at all levels.

Emphasis should be placed on teaching human rights, ROE, CA functions, negotiating skills, and inter-agency training.

NOTES

1. Charles E. Heller, *Twenty-First Century Force: A Federal Army and A Militia*, (Carlisle Barracks, PA, Strategic Studies Institute, US Army War College, 1993), p. 64.
2. See *Reserve Component Programs, Fiscal Year 1990: Report of the Reserve forces Policy Board to the President and Congress*, Office of the Secretary of Defense, 2 March 1991, p. 39.
3. For the 20 CA functional areas, see n. 15 to Chapter 2.
4. See discussion of CA missions in Barnes, 'Civil Affairs: Diplomat Warriors in Contemporary Conflict', *Special Warfare* (Winter 1991), p. 4.
5. Carnes Lord, Project Director, *Civil Affairs: Perspectives and Prospectus* (draft, February 1993), Institute for National Strategic Studies, National Defense University, p. 9–12 (hereinafter *CA Perspectives*).
6. Wayne A. Downing, 'Civil Affairs Wins the Peace', letter to the editor, *Military Review* (February 1994), pp. 3, 64.
7. George A. Joulwan, 'Operations Other Than War: A CINC's Perspective', *Military Review* (February 1994), p. 5.
8. Editorial by Jim Hoagland, 'Refugees Drive Post-Cold War Politics', *The State* (Columbia, SC), 4 May 1993, p. 11A.
9. Jim Hoagland, 'Using Military for Disaster Relief', *The State* (Columbia, SC), 17 May 1991, 14–A. Jim Hoagland's proposal for military disaster relief was endorsed by James Walsh of *Time*, who suggested a multinational rapid-deployment disaster relief force headed by Japan, which he thought might be more acceptable to Third World countries suspicious of American military actions (Walsh, 'There Must Be a Better Way', *Time*, 27 May 1991, p. 33). For legal implications, see Barnes, 'Civic Action, Humanitarian Assistance, and Disaster Relief: Military Priorities in Low-Intensity Conflict', *Special Warfare* (Fall 1989), p. 34.
10. Andrew S. Natsios, 'The International Humanitarian Response System', *Parameters* (Spring 1995), pp. 68, 79.
11. Following Hurricane Hugo in South Carolina and the California earthquake in 1989, Mayor Riley of Charleston, SC, Congressman Arthur Ravenel, Jr., of South Carolina and Congressman Leon Panetta of California all agreed that FEMA (the civilian agency in charge of domestic disaster relief) had made a mess of things and that the Army should be given primary responsibility for disaster relief. See Lee Bandy, 'FEMA's Mismanagement A Disaster, Critics Say', *The State* (Columbia, SC), 2 May 1990, p. 1–A. Hurricane Andrew in 1992 confirmed the need for a larger military role in domestic disaster relief. See editorials in *The State*, 'Army's Role Essential In Major Disaster Areas', 3 September 1992, p. 10–A, and 'Andrew Underscored New Role for Military', 23 September 1992, 10–A. As to floods, see David Evans of *The Chicago Tribune*, 'Military AWOL from Flood Crisis', reprinted in *The State*, 9 August 1993, p. 9–A.
12. Heller, n.1 supra, at pp. 65–66, 79.
13. Legal and regulatory restrictions on the use of federal troops to assist state

disaster relief operations, the lack of uniform DOD command and control structures, and layers of federal bureaucracy within FEMA make it all but impossible for Army reservists to assist Army National Guard units on an emergency basis. See 42 USC 5121 *et seq.*; DOD Directive 3025.1 (*Use of Military Resources During Peacetime Civil Emergencies Within the US*); and AR 500–6; (*Disaster Relief*).

14. For the 20 CA functional specialty areas see n. 15 in Chapter 2.
15. The complex laws and regulations that make it difficult and time consuming for governors (who command STARCS) to obtain the assistance of federal forces in a domestic disaster (see n. 13, supra) also make it impossible for a governor to command such federal forces. Integrated command occurs only when the President declares a national emergency and federalizes the National Guard, effectively removing them from the command of the governor and placing them under the control of the federal forces deployed to provide emergency assistance.
16. Heller, n. 1 supra, p. 79.
17. Ibid. at p. 65.
18. For references to The Posse Comitatus Act see n. 34 to Chapter 2.
19. The Air National Guard is creating a CA capability in the STARC; like the Marines (see note 32, infra), it has recognized the close relationship between the SJA and the CMO and is training its senior legal advisor to provide CA as well as legal support. *CA Perspectives*, n. 5 supra, supports the need for CA personnel to assist civil authorities in domestic emergencies, and the need for 'positive co-ordination' with the STARCs. After noting that '. . . the time may have come for a broad reconsideration of DOD involvement in such events', the study recommended 'A feasibility study should be conducted of the potential for utilization of USAR CA personnel and units in support of the national guard in domestic disasters' (pp. 31 and 32).
20. Charles Heller has recommended a mix of reserves, using the overseas deployment training program (ODT), and active component forces for a forward presence; see Heller, n. 1 supra at pp. 70–71.
21. See John B. Haseman, 'The FAO: Soldier-Diplomat for the New World Order', *Military Review* (September 1994), p. 74. The importance of the FAO in ethnic conflict has been emphasized by William A. Stofft and Gary L. Guertner in 'Ethnic Conflict: The Perils of Military Intervention', *Parameters* (Spring 1995), pp. 30, 40.
22. The ten Special Operations Activities assigned to USSOCOM by 19 USC 167(j) are: direct action, strategic reconnaissance, unconventional warfare, foreign internal defense, civil affairs, psychological operations, counterterrorism, humanitarian assistance, theater search and rescue, and 'other activities'. All but direct action and strategic reconnaissance activities involve political objectives that require public support. These special operations missions are discussed in JCS PUB 3–05, Chap. 2.
23. Downing, no. 6 supra at p. 63.
24. *CA Perspectives*, no. 5 supra, at 12–13.
25. There have been other proposals to provide inter-agency educational opportunities for diplomat warriors, including a LIC 'schoolhouse' at the Foreign Service Institute or National Defense University. See William J. Olson, 'Organizational Requirements for LIO', *Military Review* (January 1988), p. 16.
26. See FM 100–20, *Military Operations in Low Intensity Conflict* (Coordinating Draft, USC&GSC, Fort Leavenworth, Kansas, January 1988), pp. 1–10. For

a discussion of the merits of a SAF in LIC, see William P. Johnson and Eugene N. Russell, 'An Army Strategy and Structure', *Military Review* (August 1986), p. 69. For a discussion of a CA brigade providing command and control of a SAF in nation-building, see Raymond E. Bell, 'To Be In Charge', *Military Review* (April 1988), p. 12. Using a modified SAF as a model for future command and control of nation-building is supported in *CA Perspectives*, n. 2 supra 5, at pp. 58–59.

27. During the time of war or declared national emergency the Secretary of the Army can call up reservists or reserve units pursuant to Section 10 USC 672 *(a)*; otherwise the Secretary can call up reservists involuntarily for a period not exceeding 15 days a year pursuant to Section 10 USC 672*(b)*. To order reservists to duty for an indefinite period during peacetime the President must exercise his 200K callup authority under Section 10 USC 673*(c)*. There is currently no legal means of involuntarily ordering reservists to active duty short of mobilization during war or national emergency.

28. Downing, n. 6 supra at p. 63.

29. For the expanded role of military lawyers as advisors on law and legitimacy, see Barnes 'Legitimacy and the Lawyer in Low Intensity Conflict: Civil Affairs Legal Support', *The Army Lawyer* (October 1988), p. 5.

30. FM 101-5, *Staff Organization and Operations*, Department of the Army (May 1984), pp. 3–11, 3–12, 3–31, 3–32.

31. JCS PUB 3–57, pp.II–3; IV–6; FM 41–10, *Civil Affairs Operations*, Department of the Army (1985), pp. 1–4, 6–10.

32. Unclassified message from the Commanding General of the Marine Corps Combat Development Command, Quantico, VA, dated 7 April 1988. The Air National Guard is also making the legal advisor in the STARC the CA advisor as well (see n. 19 supra).

33. See Ted B. Borek, 'Legal Services During War', 120 *Military Law Review* 19, 35–40 (1988); cited in Barnes, 'Legitimacy and the Lawyer in Low Intensity Conflict: Civil Affairs Legal Support', *The Army Lawyer* (October 1988), pp. 8–9.

34. See Jeffrey F. Addicott and Andrew M. Warner, 'JAG Corps Poised for New Defense Missions: Human Rights Training in Peru', *The Army Lawyer*, February 1993, p. 78; also Addicott and Warner, 'Promoting the Rule of Law and Human Rights', *Military Review* (August 1994), p. 38.

35. Jeffrey F. Addicott and Andrew M. Warner, 'JAG Corps Poised for New Defense Missions: Human Rights Training in Peru' *The Army Lawyer* (February 1993), p. 78.

36. Joulwan, n. 7 supra, at p. 6.

37. 'International Law Note: The Role of the Military in Emerging Democracies', *The Army Lawyer* (December 1992), pp. 28.

38. Ibid. at p. 30.

39. *CA Perspectives*, n. 5 supra, pp. 23–24.

40. See 'Enforcement: command influence and military justice', in Chapter 5, and nn. 45–52 to Chapter 5. Judge advocates in the US Air Force Reserve are already participating in courts-martial as trial and defense counsel and judges.

41. S. L. Arnold and David T. Stahl, 'A Power Projection Army in Operations Other Than War', *Parameters* (Winter 1993–94), pp. 13–15.

42. Army Regulations 27–10, *Military Justice* (December 1989), para. 19–7(c).

43. See *Student Text, Military Law and Justice* (Required Readings in Military Science IV, Military Qualifications Standards I, Precommissioning Require-

ments, June 1992). The introduction to the text describes military law as it relates to the legitimacy and leadership:

> Military law has two major purposes. First, it provides the minimum standards for determining the legitimacy of military operations and activities. Second, it provides the standards, rules, and procedures for maintaining good order and discipline in the armed forces. Part I of this Student Text, Military Law, begins with the US Constitution as the foundation of military law and justice, and then examines the standards of conduct required for military legitimacy in war and peace. Part II, Military Justice, examines the standards, rules, and procedures that are required to maintain good order and discipline in our armed forces.
>
> Military operations are an extension of the political process. Like all government activities in a democracy they must be perceived as legitimate to be effective. Legitimacy is measured by public perceptions of moral authority – authority that must rest upon the rule of law. But the law can never be a substitute for moral choice. Values such as those in the Professional Army Ethic are part of the fabric of legitimacy. Law and values provide the context for the tough discipline that must be made in an often unforgiving world.
>
> This Student Text provides more than the standards, rules, and procedures that constitute military law and justice. It also provides historic background and practical applications to help you understand the spirit as well as the letter of the law. That spirit is embodied in the fundamental principles of human rights, democracy, and the rule of law – principles which are enshrined in the US Constitution. The Constitution and the values it represents are the bedrock of military legitimacy.
>
> As an Army officer you must understand the rule of law and its relationship to the Professional Army Ethic. Your oath of office requires that your ultimate loyalty be to the Constitution, a duty that requires an understanding of the military as both a shield and a sword for Constitutional rights. This text will help you relate the rule of law to the traditional military values of loyalty, duty, integrity, and selfless service. It will help you understand the meaning of legitimacy in the profession of arms.

44. Charles Heller has advocated USARF Schools supporting ROTC; see Heller, n. 1 supra, at p. 61.
45. See William Hagan, 'The Officer Corps: Unduly Distant From Military Justice?', *Military Review* (April 1991), p. 51. Colonel Hagan has argued that a lack of understanding (of the proper role of command influence) is a large part of the reason why commanders are frustrated with the complexities of military justice. To help alleviate the problem Colonel Hagan has suggested that ROTC cadets and OCS officer candidates should receive military justice training commensurate with that provided at the United States Military Academy at West Point. The 'Student Text' mentioned at n. 43 supra was prepared with that purpose in mind, as well as to relate the law to legitimacy and leadership, as set forth in n. 43. supra.
46. General McCaffrey covered these points in his keynote address to the School of the Americas on 10 August 1994, *The National Armed Forces as Supporters of Human Rights* (see references at n. 55 to Chapter 1, and nn. 57–60 to Chapter 4 supra). He completed his address with the following observation on civil-military relations and military legitimacy (without using the term): 'Our experience has been that our citizens are supportive of the armed forces if they think highly of us. How do they form their impressions of us? They form them when their sons and daughters – our soldiers, sailors, airmen, and Marines – go home and tell their families and friends that they are treated

well while they serve. They form them every time they come in contact with the armed forces: when they see a soldier travelling on leave; when they see a military convoy; and when they live beside a military base. Finally, they form them when they see us in action in a conflict or in a peaceful mission.' (McCaffrey at p. 12).

47. William A. Stofft and Gary L. Guertner, 'Ethnic Conflict: The Perils of Military Intervention', *Parameter* (Spring 1995), pp. 30, 41.
48. *CA Perspectives*, n. 5 supra, at p. 16.
49. S. L. Arnold and David T. Stahl, *A Power Projection Army in Operations Other Than War*', n. 41 supra, at p. 4, 8–12, 15–17; see also Arnold, 'Somalia: An Operation Other Than War', *Military Review* (December 1993), p. 26.
50. Ibid., Arnold, *A Power Projection Army in Operations Other Than War*, at p. 22.

Conclusion

There is a season for everything ...
A time for war, a time for peace.

Ecclesiastes 3: 1, 8

Military legitimacy reflects what is expected of the military. Its requirements provide the moral authority for the military to act as an instrument of national power. Military legitimacy represents the balance between might and right, a balance that varies according to the season: it is different in times of war and peace.

The uncertain and violent peace following the end of the Cold War requires military capabilities that are as constructive during peacetime as they are destructive during wartime. They are the capabilities needed for operations other than war, and their legitimacy is inextricably bound with public perceptions of democracy, human rights, and the rule of law.

Capabilities for operations other than war depend upon unique leadership and extensive civil-military relations to achieve military legitimacy; and painful lessons have taught that traditional combat capabilities are unsuited for these non-combat operations. The *diplomat warrior* and *civil affairs* offer new models of leadership and civil-military relations needed for the new millennium. They complement and do not degrade warfighting capabilities, providing more effective utilization of existing combat support and service support forces.

Creating the military capabilities and leadership needed for operations other than war will require changes in the world's largest bureaucracy, which is noted for its resistance to change. For many traditionalists in the Pentagon any joint venture with civilian agencies is a dangerous diversion for professional warriors. But if the Army is to be all that it can (and must) be in the new millennium, its leaders must be equally at home in both civilian and military environments.

The needs of the new strategic environment call for a new generation

187

of diplomat warriors whose concept of professionalism is grounded in military legitimacy. In a seamless Total Force civilian soldiers must be accepted as full partners – not just reservists – serving with their full-time counterparts as an extension of both the US military and diplomatic corps.

Select Bibliography

Jeffrey F. Addicott, 'Operation Desert Storm: R. E. Lee or W. T. Sherman?', *Military Law Review*, Volume 136 (Spring 1992), pp. 115, 133.

—, and Andrew M. Warner, 'JAG Corps Poised for New Defense Missions: Human Rights Training in Peru', *The Army Lawyer* (Feb. 1993), p. 78.

—, and Andrew M. Warner, 'Promoting the Rule of Law and Human Rights, *Military Review* (Aug. 1994), p. 38.

Maxwell Alston, 'Military Support to Civil Authorities: New Dimensions for the 1990s', *The Officer* (Oct. 1991), p. 28.

S. L. Arnold and David T. Stahl, 'A Power Projection Army in Operations Other Than War', *Parameters* (Winter 1993–94), pp. 13–15.

A. J. Bacevich, 'New Rules: Modern War and Military Professionalism', *Parameters* (Dec. 1990), p. 12.

Rudolph C. Barnes, Jr., 'Special Operations and the Law', *Military Review* (Jan. 1986), p. 49.

—, 'Civil Affairs: A LIC Priority', *Military Review* (Sept. 1988), p. 38.

—, 'Legitimacy and the Lawyer in Low-Intensity Conflict (LIC): Civil Affairs Legal Support', *The Army Lawyer* (Oct. 1988), p. 5.

—, 'Civic Action, Humanitarian and Civic Assistance, and Disaster Relief: Military Priorities in Low Intensity Conflict', *Special Warfare* (Fall 1989), p. 34.

—, 'The Diplomat Warrior', *Military Review* (May 1990), p. 55.

—, 'Civil Affairs: Diplomat-Warriors in Contemporary Conflict', *Special Warfare* (Winter 1991), p. 4.

—, 'Military Legitimacy and the Diplomat Warrior', *Small Wars and Insurgency* (Spring/Summer 1993), p. 1.

Raymond E. Bell, 'To Be In Charge', *Military Review* (April 1988), p. 12.

William R. Berkman, 'Civil Affairs in Vietnam', unpublished paper written for US Army War College, Carlisle Barracks, PA, Dec. 1973.

Ted B. Borek, 'Legal Services During War', 120 *Military Law Review*, 1988, p. 35.

Boutros Boutros-Ghali, 'Empowering the United Nations', *Foreign Affairs* (Winter 1992–93), pp. 89, 94.

James K. Bruton and Wayne D. Zajac, 'Cultural Interaction: The Forgotten Dimension of Low Intensity Conflict', *Special Warfare* (April 1988), p. 29.

Stephen P. Bucci, 'Fighters vs. Thinkers: The Special Operations Staff Officer Course and the Future of SOF', *Special Warfare* (Spring 1989), p. 33.

Philip Caputo, *A Rumor of War* (New York: Ballantine Books, 1986).

Carl von Clausewitz, *On War*, edited and translated by Michael Howard and Peter Paret (Princeton, NJ: Princeton University Press, 1984).

Dennis F. Coupe, 'Commanders, Staff Judge Advocates, and the Army Client', *The Army Lawyer* (Nov. 1989), p. 3.

Barry D. Crane, et al., 'Between Peace and War: Comprehending Low Intensity Conflict', *Special Warfare* (Summer 1989), p. 2.

Martin von Creveld, *The Transformation of War* (New York: The Free Press).

Robert Cullen, 'The Human Rights Quandary', *Foreign Affairs* (Winter 1992–93), p. 79.

Cecil B. Currey, 'Edward G. Landsdale: LIC and the Ugly American', *Military Review* (May 1988), p. 50.

James D. Delk, 'Military Assistance in Los Angeles', *Military Review* (Sept. 1992), p. 13.

Geoffrey Demarest, 'Updating the Geneva Conventions: The 1977 Protocols', *The Army Lawyer* (Nov. 1983), p. 18.

David Donovan, *Once A Warrior King* (New York: Ballantine Books, 1985).

Thomas R. Dubois, 'The Weinberger Doctrine and the Liberation of Kuwait', *Parameters* (Winter 1991–92), p. 24.

Charles J. Dunlap, Jr., 'The Origins of the American Military Coup of 2012', *Parameters* (Winter 1992–92), p. 2.

Jonathan T. Dwarken, 'Rules of Engagement: Lessons from Restore Hope', *Military Review* (Sept. 1994), p. 26.

Stephen Dycus, Arthur L. Berney, William C. Banks and Peter Raven-Hansen, *National Security Law* (Boston: Little, Brown and Company, 1990).

Thomas K. Emswiler, 'Security Assistance and Operations Law', *The Military Lawyer* (Nov. 1991), p. 10.

Richard J. Erickson, *Legitimate Use of Military Force Against State-Sponsored International Terrorism* (Maxwell AFB, Alabama: Air University Press, 1989).

John T. Fischel and Edmund S. Cowan, 'Civil-Military Operations and the War for Moral Legitimacy in Latin America', *Military Review* (Jan. 1988), p. 41.

Shelby Foote, *The Civil War, Volume 2: Fredericksburg to Meridian* (New York: Random House, 1986).

Alexander A. C. Gerry, 'Former USSR Asks Reserve Officer Help for Ex-Officer, ROA National Security Report', *The Officer* (Dec. 1992), p. 23.

Robert L. Goldrich and Stephen Daggett, 'Defense Policy: Threats, Force Structure, and Budget Issues', Foreign Affairs and National Defense Division (IB90013) 15 Dec. 1992.

Gidon Gottlieb, *Nation Against State* (New York: Council on Foreign Relations Press, 1993).

David H. Hackworth, *About Face* (New York: Simon & Schuster, 1989).

William Hagan, 'The Officer Corps: Unduly Distant From Military Justice?' *Military Review* (April 1991), p. 51.

David L. Hall, 'The Constitution and Presidential War Making Against Libya', *Naval War College Review* (Summer 1989), p. 76.

B. H. Liddel Hart, 'National Object and Military Aim', *Strategy*, 2nd rev. ed. (New York: Frederick A. Praeger, 1967).

Anthony E. Hartle, *Moral Issues in Military Decision Making* (Lawrence, Kansas: Univeristy Press of Kansas, 1989).

John B. Haseman, 'The FAO: Soldier-Diplomat for the New World Order', *Military Review* (Sept. 1994), p. 74.

Charles E. Heller, *Twenty-First Century Force: A Federal Army and Militia*, Strategic Studies Institute, US Army War College, Carlisle Barracks, PA (1993).

—, 'A Cadre System for the US Army', *Military Review*, Oct. 1991, p. 14.

John Hersey, *Hiroshima* (New York: Bantam Books, 1986).

John B. Hunt, 'Hostilities Short of War', *Military Review* (March 1993), p. 41.

Samuel P. Huntington, 'The Clash of Civilizations', *Foreign Affairs* (Summer 1993), p. 22.

—, *The Soldier and the State* (Cambridge, MA: The Belknap Press of Harvard University Press, 1957).

Thomas Jefferson, *The Jefferson Bible* (New York: Clarkson N. Potter, Inc., 1964).

William P. Johnson and Eugene N. Russell, 'An Army Strategy and Structure', *Military Review* (Aug. 1986), p. 69.

David A. Jonas, 'Fraternization: Time for a Rational Department of Defense Standard', *Military Law Review* (Winter 1992), p. 37.

George A. Joulwan, 'Operations Other Than War: A CINC's

Perspective', *Military Review* (Feb. 1994), p. 5.

Robert D. Kaplan, 'The Coming Anarchy', *The Atlantic Monthly*, Feb. 1994, p. 44.

Steven Keeva, 'Lawyers in the War Room', *The ABA Journal* (Dec. 1991), p. 52.

Paul Kennedy, *The Rise and Fall of the Great Powers* (New York: Random House, 1987).

Cole C. Kingseed, 'Peacetime Engagement: Devising the Army's Role', *Parameters* (Autumn 1992), p. 98.

Jacob W. Kipp, 'Civil-Military Relations in Central and Eastern Europe', *Military Review* (Dec. 1992), p. 27.

Henry Kissinger, *Diplomacy* (New York: Simon & Schuster 1994).

Michael T. Klare, 'The Interventionist Impulse: US Military Doctrine for Low Intensity Warfare', in Peter Kornbluh's *Low Intensity Warfare* (New York: Pantheon Books, 1988).

Richard H. Kohn, 'Women in Combat, Homosexuals in Uniform: The Challenge of Military Leadership', *Parameters* (Spring 1993), p. 2.

Lewis H. Lapham, 'Notebook: God's Gunboats', *Harper's Magazine*, Feb. 1993, p. 10.

Walter Laqueur, 'Russian Nationalism', *Foreign Affairs* (Winter 1992–93), p. 103.

William Lederer and Eugene Burdick, *The Ugly American* (New York: W. W. North & Co. 1958).

James R. Locher, III, 'Low Intensity Conflict: Challenges for the 1990s', *Defense '91* (July/August 1991), p. 19.

John Lukacs, 'The End of the Twentieth Century', *Harpers* (Jan. 1993), p. 39.

Robert L. Maginnis, 'A Chasm of Values', *Military Review* (Feb. 1993), pp. 2–11.

Charles S. Maier, 'Democracy and Its Discontents', *Foreign Affairs* (July/Aug. 1994), p. 48.

Gabriel Marcella, 'The Latin American Military, Low Intensity Conflict, and Democracy', *Winning the Peace: The Strategic Implications of Military Civic Action*, eds. John W. De Pauw and George A. Luz (Strategic Studies Institute, US Army War College, Carlisle, PA), Ch. 4.

Mark S. Martins, 'Rules of Engagement for Land Forces: A Matter of Training, Not Lawyering', *Military Law Review*, Vol. 143 (Winter 1994), p. 3.

Peter Maslowski, 'Army Values and American Values', *Military Review* (April 1990), p. 10.

Lloyd J. Matthews, 'Is the Military Profession Legitimate?', *Army* (Jan. 1994), p. 15.

Steven Metz, *America in the Third World* (Strategic Studies Institute, US Army War College, Carlisle Barracks, PA, 1994).

—, *The Future of Insurgency* (Strategic Studies Institute , US Army War College, Carlisle Barracks, PA, 1993).

Andrew S. Notsios, 'The International Humanitarian Response System', *Parameters* (Spring 1995), p. 68.

Sam Nunn, 'The Fundamental Principles of the Supreme Court's Jurisprudence in Military Cases', *The Army Lawyer* (Jan. 1995), p. 27.

William V. O'Brian: 'Special Operations in the 1980s: American Moral, Legal, Political, and Cultural Constraints', *Special Operations in US Strategy* (National Defense University Press, 1984), pp. 53, 58; and unpublished paper entitled 'Just War Doctrine's Complementary Role in the International Law of War' (1991).

William J. Olson, 'Organizational Requirements for LIC', *Military Review* (Jan. 1988), p. 16.

David R. Palmer, 'The Constitution and the US Army: The Commander in Chief', *The Constitution and the US Army* (US Army War College and US Army History Institute, Carlisle Barracks, PA, 1988), p. 76.

John Embry Parkinson, Jr., 'United States Compliance with Humanitarian Law Respecting Civilians During Operation Just Cause', *Military Law Review* (Summer 1991), p. 31.

W. Hays Parks, 'Teaching the Law of War', *The Army Lawyer* (June 1987), p. 4.

David H. Petraeus, 'The Just War Tradition', *Military Review* (April 1984), p. 84.

Colin L. Powell, 'US Forces: Challenges Ahead', *Foreign Affairs* (Winter 1992–93), p. 32.

William H. Riley, Jr., 'Challenges of a Military Advisor', *Military Review* (Nov. 1988), p. 34.

John Gerard Ruggie, 'Wandering the Void: Charting the UN's Strategic Role', *Foreign Affairs* (Nov./Dec. 1993), p. 26.

Stanley Sadler, 'Seal the Victory: A History of US Army Civil Affairs', *Special Warfare* (Winter 1991), p. 38.

Sam C. Sarkesian, 'Organizational Strategy and Low Intensity Conflicts', *Special Operations in US Strategy*, eds. Frank R. Barnett, B. Hugh Tovar and Richard H. Shultz (New York: National Defense University Press, National Strategic Information Center, Inc., 1984).

Leon V. Segal, 'The Last Cold War Election', *Foreign Affairs* (Winter 1992–93), p. 8.

Charles M. Simpson, III, *Inside the Green Berets: The First Thirty*

Years: A History of the US Army Special Forces (Novato, CA: Presidio Press, 1983).

Abraham D. Sofaer, 'Terrorism, the Law, and the National Defense', *Special Warfare*, Fall 1989, p. 12.

John E. Shephard, 'Thomas Becket, Ollie North, and You: The Importance of an Ethical Command Climate', *Military Review* (May 1991), p. 21.

Ranier H. Spencer, 'A Just War Primer', *Military Review* (Feb. 1993), p. 20.

Shelby L. Stanton, *The Rise and Fall of an American Army* (New York: Dell Publishing Company, Inc., 1988).

Stephen John Stedman, 'The New Interventionists', *Foreign Affairs* (Winter 1992–93), Vol. 72, No. 1, p. 1.

William A. Stofft and Gary L. Guertner, 'Ethnic Conflict: The Perils of Military Intervention', *Parameters* (Spring 1995), p. 30.

Gordon R. Sullivan, 'The Challenges of Peace', *Parameters* (Autumn 1994), p. 4.

Harry G. Summers, Jr., *On Strategy: the Vietnam War in Context* (Strategic Studies Institute, US Army War College, Carlisle Barracks, PA), 1989.

Richard E. Taylor and John D. McDowell, 'Low Intensity Campaigns', *Military Review* (March 1988), p. 2.

Barbara W. Tuchman, *A Distant Mirror* (New York: Ballantine Books, 1978).

UN General Assembly (Forty-seventh session) and Security Council: Report of the Secretary-General on the Work of the Organization (17 June 1992), *An Agenda for Peace*.

UN General Assembly (Fiftieth session) and Security Council: Report of the Secretary-General on the Work of the Organization (3 Jan. 1995), *Supplement to An Agenda for Peace*.

UN High Commissioner for Refugees, *Handbook for Emergencies*, Geneva, 1982.

Kurt Vonnegut, *Slaughterhouse Five* (New York: Dell Publishing Company Inc., 1968).

Malham M. Wakin, *War, Morality, and the Military Profession* (Boulder, Colorado: Frederick A. Praeger, 1986).

Fran W. Walterhouse, 'Using Humanitarian Activities as a Force Multiplier and a Means of Promoting Stability in Developing Countries', *The Army Lawyer* (Jan. 1993), p. 16.

Harry F. Walterhouse, *A Time to Build* (Columbia, SC: The R. L. Bryan Company, University of South Carolina Press, 1964).

Michael Walzer, *Just and Unjust Wars* (New York: Basic Books, 1977).

Ralph R. Young, 'Snapshots of Civil Affairs: A Historical Perspective and Views', unpublished paper presented at the 39th Annual

Conference of the Civil Affairs Association at San Antonio, Texas, June 1986.

Yuval Joseph Zacks, 'Operation Desert Storm, A Just War?', *Military Review* (Jan. 1992), p. 20.

GOVERNMENT AND MILITARY PUBLICATIONS

A National Security Strategy of Engagement and Enlargement, July 1994, The White House, Washington, DC.

A National Military Strategy for the US, February 1995, Department of Defense, Washington, DC.

The Lieber Code, General Order Number 100, dated 24 April 1863, published in *The Military Laws of the United States*, War Department Document No. 64 (Washington, Government Printing Office, 1897), pp. 779–99.

The Manual for Courts Martial (MCM), 1984 (as amended).

Standards of Ethical Conduct For Employees of the Executive Branch, including Part I of Executive Order 12674 and 5 C.F.R. Part 2635 Regulation, prepared by the US Office of Government Ethics, Washington, DC, Aug. 1992, Part I.

Report of the US General Accounting Office (GAO) to the Chairman, Subcommittee on Military Personnel and Compensation, Committee on Armed Services, House of Representatives, entitled *Army Force Structure: Future Reserve Roles Shaped by New Strategy, Base Force Mandates, and Gulf War* (B-251524), Dec. 1992.

Rand Report, *Assessing the Structure and Mix of Future Active and Reserve Forces: Final Report to the Secretary of Defense* (ISBN: 0-9330-1299-1) RAND, 1992.

Reserve Component Programs, Fiscal Year 1990: Report of the Reserve Forces Policy Board to the President and Congress, Office of the Secretary of Defense, 2 March 1991.

Civil Affairs in the Persian Gulf War, Symposium Proceedings, held 25–27 Oct. 1991 at US Army John F. Kennedy Special Warfare Center and School, Fort Bragg, NC.

Civil Affairs: Perspectives and Prospects (draft, Feb. 1993), Carnes Lord, Project Director, Institute for National Strategic Studies, National Defense University, pp. 6–8.

DOD Directory of Military and Associated Terms (1 Dec. 1989).

DOD Directive No. 2000.1, 1972, and AR 550–1, 1981, both of which are entitled *Procedures for Handling Requests for Political Asylum and Temporary Refuge*.

DOD Directive No. 2205.2, 6 Oct. 1994, *Humanitarian and Civic Assistance (HCA) Provided in Conjunction with Military Operations*.

DOD Directive 3025.1, *Use of Military Resources During Peacetime Civil Emergencies Within the US*.

DOD Directive 5100.77, *The DOD Law of War Program* (July 1979).

DOD Instruction No. 2205.3, 27 Jan. 1995, *Implementing Procedures for the Humanitarian and Civic Assistance (HCA) Program*.

JCS (Test) PUB 3–0, *Doctrine for Unified and Joint Operations* (Jan. 1990).

JOINT PUB 3–05, *Doctrine for Joint Special Operations* (draft, Jan. 1994).

JCS PUB 3–07, *Doctrine for Joint Operations in Low Intensity Conflict*, The Joint Chiefs of Staff (Final Draft, Jan. 1990).

JCS PUB 3–57, *Joint Civil Affairs Operations* (Final Draft, Nov. 1990).

FM 22–100, *Military Leadership* (Co-ordinating Draft, 15 June 1988).

FM 27–1, *Legal Guide for Commanders*, 1987.

FM 27–10, *The Law of Land Warfare* (July 1956, including Change 1 dated July 1976).

FM 27–100, *Legal Operations*, Sept. 1991.

FM 41–10, *Civil Affairs Operations*, Dec. 1985.

FM 100–1, *The Army* (May 1986).

FM 100–5, *Operations*, Headquarters, Department of the Army, June 1993.

FM 100–20 (AFP 3–20), *Military Operations in Low Intensity Conflict*, Headquarters, Department of the Army and Department of the Air Force, 1 Dec. 1989.

FM 100–23, *Peace Operations* (draft), Headquarters. Department of the Army, 26 Jan. 1994.

FM 100–25, *Doctrine For Army Special Operations Forces*, Department of the Army (Final Draft, Oct. 1990).

AR 27–10, *Army Legal Services*, 1989.

AR 380–67, *Personnel Security Program* (9 Sept. 1988).

AR 500–60, *Disaster Relief*.

AR 600–20, *Army Command Policy*, 1988.

AR Pam 27–21, *Administrative and Civil Law Handbook*, 1992.

Draft *Operational Law Handbook* (JA 422, 1993) prepared by the Center for Law and Military Operations and the International Law Division at The Judge Advocate General's School, US Army, at Charlottesville, Virginia.

The Human Rights Deskbook, United States Army Special Forces Command Center for Special Operations Law, Fort Bragg, NC.

Student Text, Military Law and Justice, Required Readings in Military Science IV, Military Qualification Standards I (Precommissioning Requirements), prepared by the Department of Law, United States Military Academy, for the US Army Cadet Command, June 1992.

Index

For Product Safety Concerns and Information please contact our EU
representative GPSR@taylorandfrancis.com
Taylor & Francis Verlag GmbH, Kaufingerstraße 24, 80331 München, Germany

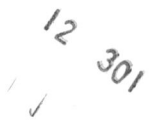